Prai...
Siddhart...

"Fascinating. . . . Weaves ancient spiritual wisdom with today's latest scientific research about mindful practices and their effects on mental health. . . . Whether you're a skeptic or a true believer, exploring Siddhartha's brain offers compelling insights and invites further questions about the potential of the human mind."
—*Chicago Tribune*

"Reveals not only how mindfulness meditation can rewire the human brain and help us achieve a sense of spiritual fulfillment but also how we can easily integrate the practice into our daily lives." —*Scientific American*

"Kingsland expertly weaves the story and teachings of the Buddha with clinical and scientific research to engage in a highly readable examination of the benefits of mindfulness and meditation. . . . Adeptly incorporates Buddhist and scientific terminology in such a way that concepts are easily understood through the context of the narrative. . . . A satisfying read."
—*Library Journal* (starred review)

"It's a pleasure to read *Siddhartha's Brain*. . . . A smart, accessible balance of philosophical teachings and brain science and how meditation can relate to everything from addiction to Alzheimer's disease." —Associated Press

"[A] fascinating exploration of the neuroscience behind meditation. . . . Kingsland skillfully dives in and out of various subjects—the neurological relaxation response to meditation, the difference between pain and suffering, emotional regulation—and effectively paints a neurological picture of the mind without devaluing Buddhism's spiritual image of cognition." —*Publishers Weekly*

"Tracing Western scientific research on the subject of meditation from the 1960s to the present day, the author details studies demonstrating the efficacy of mindfulness in battling chronic pain, mental illness, addiction, and even aging while emphasizing that the benefits of mindfulness have been enjoyed for centuries without the help of modern science. . . . Brain science and Buddhist lore combine in this compelling treatise on the benefits of meditation and mindfulness." —*Kirkus Reviews*

SIDDHĀRTHA'S
BRAIN

SIDDHĀRTHA'S BRAIN

UNLOCKING THE ANCIENT
SCIENCE OF ENLIGHTENMENT

JAMES KINGSLAND

WILLIAM MORROW
An Imprint of HarperCollinsPublishers

Figures, unless otherwise credited, provided courtesy of the author.

Map by Narasit Nuad-o-Lo.

Figure 2. *Top: Patrick J. Lynch (Wikimedia Commons, adapted); bottom:* Sobotta's Anatomy Atlas, *1908 edition (adapted).*

Figure 3. *Patrick J. Lynch (Wikimedia Commons, adapted).*

Figure 4. *Patrick J. Lynch (Wikimedia Commons, adapted).*

Figure 5. *Patrick J. Lynch (Wikimedia Commons, adapted).*

Figure 6. *Patrick J. Lynch (Wikimedia Commons, adapted).*

Figure 7. *Patrick J. Lynch (Wikimedia Commons, adapted).*

HarperCollins books may be purchased for educational, business, or sales promotional use. For information, please e-mail the Special Markets Department at SPsales@harpercollins.com.

A hardcover edition of this book was published in 2016 by William Morrow, an imprint of HarperCollins Publishers.

FIRST WILLIAM MORROW PAPERBACK EDITION PUBLISHED 2017.

Library of Congress Cataloging-in-Publication Data has been applied for.

ISBN 978-0-06-240387-2

HB 04.05.2023

CONTENTS

INTRODUCTION

"We are all mentally ill," said the smiling monk in the wide-brimmed hat, as if this explained everything. My partner and I were staying a couple of nights as guests at Amaravati Buddhist Monastery, near Hemel Hempstead in the Chiltern Hills of southern England. I was a science journalist with the *Guardian* newspaper and had traveled up by train from London the previous day to interview the abbot, a kindly Englishman in his fifties named Ajahn Amaro, who has been trained in the strict Thai Forest tradition of Buddhism. The three of us stood in bright morning sunshine on a path that led between neat flower beds from the painted wooden huts of the monastery's retreat center to a field of rough grass, where men and women were pacing very slowly and deliberately, each absorbed in a private world of his or her own. Some were walking back and forth between trees, following tracks worn in the grass by thousands of tramping feet. Others were relentlessly circling a bell-shaped, granite stupa at the center of the field.

A two-week retreat for about thirty laypersons had begun the previous evening, and this morning the abbot—the monk in the sun hat—had sent them out into the grounds to practice walking meditation. His observation about our collective neurosis took me by surprise, following as it did from my own observation that the otherworldly activity in the field reminded me of a scene from a zombie movie I once saw. On reflection, it wasn't the most enlightened comment to direct to a revered Buddhist teacher or *ajahn* during a meditation retreat, but I was tired and grouchy after being awakened at four thirty in the morning by the monastery's great brass bell being struck somewhere outside in the darkness, summoning us from our dormitory to the meditation hall for an hour of chanting and contemplation.

I only later discovered that in Buddhist philosophy, a human being is not considered completely sane until he or she has become fully enlightened.[1] Buddhists believe the mechanism of the human mind is faulty, like a clock running too fast or too slow. No matter how rational or mentally fit we believe ourselves to be, much of our lives is spent obsessing about our social and professional standing, about getting sick and growing old, yearning for more of this and less of that, chewing over our faults and those of other people. Buddhists believe that our minds create *dukkha*: the suffering or sense of "unsatisfactoriness" that is part and parcel of ordinary human existence, the incessant itch of wanting more pleasure and more possessions, trying to hold on to some experiences while frantically trying to push others away. To observe that everybody is mentally ill was the monk's way of summing up this shared psychological predicament.

Earlier that day, in the thin gray light before dawn, sitting cross-legged on the floor with the monks and nuns before the gilded Buddha in the monastery's meditation hall, we had chanted:

Birth is dukkha;
Aging is dukkha;
Death is dukkha;

Sorrow, lamentation, pain, grief, and despair are dukkha;
Association with the disliked is dukkha;
Separation from the liked is dukkha;
Not attaining one's wishes is dukkha.

This was a far cry from the brisk, cheerful hymns we used to sing at morning assembly in the chapel of the Methodist boarding school I attended as a child. Instead of affirming the triumph of celestial beings over evil, here was a stark reminder that all human existence is mired in suffering. The message seemed to be that no one gets to live happily ever after: everything was *not* going to be okay. Regardless of the joys, loves, and achievements scattered along life's path, around every corner awaited loss, disappointment, sickness, aging, and death. There would be no escaping these things, no matter how hard we worked, how much we earned, how healthily we ate, how often we went to the gym. It was an ancient formulation of the modern refrain, "Life is a bitch, and then you die."

You may find this sort of reflection needlessly maudlin or you may see it as a bracing admission of the truth. Speaking for myself, I found the sentiment liberating. By saying the words out loud, we were acknowledging the lies we continually tell ourselves to get through the day. The down-to-earth honesty of the chant moved me. It brought a sense of reconciliation with reality. Even so, I was startled by the monk's assertion that "we are all mentally ill." Surely it's one thing to suffer because of one's circumstances—loss, failure, ill health, aging—and quite another to experience an unrelenting illness such as major depression or psychosis, conditions that are always there in the shadows regardless of how well or badly things are going? Surely these illnesses fall into a different category of dukkha that is experienced only by an unfortunate few?

This view is beginning to look increasingly simplistic. We have become accustomed to the idea that there are two kinds of people: those who suffer from a psychiatric illness and those with a clean bill of mental health. In reality, the picture is much more blurred. Psychiatrists are beginning to realize that traditional diagnoses such as depression, anxiety, schizophrenia, and bipolar disorder are not as clear-cut as they once believed them to be, and that symptoms used to label patients as having one illness or another are in fact widespread and exist on a continuum in the general population.[2, 3]

Take psychosis, a condition popularly assumed to be extremely rare. It is traditionally characterized as the experience of confused, disturbing thoughts, hallucinations, and delusions such as paranoia (the unfounded belief that other people are trying to harm us). In reality, hallucinations and paranoia are much more prevalent in the general population than many realize. Research suggests that up to 30 percent of us will have daytime hallucinatory experiences sometime during our lives, and somewhere between 20 and 40 percent are regularly prey to paranoid thoughts.[4-6] Even among those who have been diagnosed as having psychosis, there is an enormous amount of variation in their experience of delusions and hallucinations. It seems that what unites people with "psychosis" more than any other symptom is their experience of anxiety, depression, and neuroticism, all of which are commonplace among people who have never been labeled as mentally ill.[7] Muddying the waters still further, patients with severe depression often experience the delusions and hallucinations traditionally associated with psychosis.

Another example is bipolar disorder, characterized by alternating bouts of depression and elation or hyperactivity. While only

1 to 1.5 percent of people in Europe and the US are diagnosed as having bipolar disorder, mood swings are commonplace, and as many as 25 percent of us report experiencing periods of euphoria, reduced need for sleep, and racing thoughts. According to the British Psychological Society, this suggests that an "all or nothing" diagnosis for bipolar disorder is an oversimplification, as it is for psychosis, and that symptoms of the disorder exist on a continuum throughout the general population.[8]

So it seems there is a background level of psychological malaise that touches both the "mentally well" and the "mentally ill." The formal diagnoses are just the tip of an iceberg, though the part of the iceberg showing above the water is quite bad enough. Mental health services have their work cut out, even in countries such as Denmark, which for years rejoiced under the title of "the happiest nation on Earth," thanks to its high GDP per capita, low income inequality, personal freedoms, good nutrition, excellent public health care, long life expectancy, and other markers.[9] Despite all their nation's blessings, a surprisingly high number of Danes require treatment for serious mental illnesses at some point. Some 38 percent of Danish women and 32 percent of Danish men will receive therapy in a psychiatric hospital or clinic during their lifetime.[10] Clearly, many others, both in Denmark and in every other country around the world, experience the symptoms of mental illness without resorting to this kind of specialist professional help. They are the silently suffering majority: the everyday mentally unwell who struggle on largely by themselves.

Mental health problems start early in life. Worldwide, around 10 percent of children are estimated to have a diagnosable mental illness, roughly half being anxiety disorders and half a conduct disorder or ADHD (Attention Deficit Hyperactivity Disorder).[11]

Many of these children will grow up unhappy. The best predictor of whether a child will become an adult who is satisfied with life is not academic achievement, sociability, or family background, but his or her emotional health during childhood.[12]

The high prevalence of psychiatric illnesses and the fact that their symptoms exist, on a continuum of severity, throughout the general population suggests that they are not discrete conditions like diabetes or asthma but an extreme manifestation of the ordinary human condition. Genetics, upbringing, and life events certainly play a powerful role in making some people more susceptible than others, but our shared mental endowment—the standard-issue cerebral equipment, if you will—is largely to blame for all this psychological turmoil. Traditional diagnoses of mental illness capture only a fraction of our problems, and the widespread prevalence of violence, prejudice, and conflict in human society are hardly indicators of well-tuned mental machinery.

What's to be done? It's not as if we haven't been trying to fix the innate weaknesses of the human mind for a very long time. Attempts to fix our wonky brains are as old as civilization. You could argue that the only common ground between the world's great religions is that they have been doing their level best for millennia to bring the wayward mind to heel. So when Ajahn Amaro asserted that "we are all mentally ill," there was an even bolder subtext: "Buddhism is the cure." All religions are trying to achieve the same objective in their own ways, with varying degrees of success. What seems to set his apart from the others is that it aims to achieve this daunting feat without a rigid creed, set of commandments, or appeals to divine intercession.

Many have argued that Buddhism is not a religion at all, at least not in the conventional sense. To an atheist and skeptic like myself, this lack of a supernatural belief system makes Buddhism very appealing. When I first became interested in its practices and philosophy, some five years ago, I was also intrigued by the way "sin" in the language of other religions—lust, gluttony, sloth, wrath, envy, pride, and so forth—is labeled more neutrally by Buddhists as "unskillful behavior" that will reap painful consequences through the ineluctable operation of the laws of cause and effect. The implication seems to be that to be a good, contented human being is a skill that can be learned, like driving a car or baking a cake. The more you practice, the better you will get at it. Viewed this way, to judge someone for their greed or pride starts to look as misguided as condemning them for not being able to drive or bake.

Nevertheless, why should Buddhism be any better than the other world religions—or indeed a completely secular approach—at teaching such skills? All things mystical and religious, regardless of whether or not they involve a god, a creed, or commandments, are viewed with suspicion by many scientists and nonbelievers, including the majority of those I have worked with over the years in my job as a science writer and editor. And the cure Buddhism claims to offer for the afflicted human mind is largely based on meditation, which for professional skeptics like myself looks initially like just another health fad. Mindfulness meditation, which involves cultivating nonjudgmental awareness of the present moment, has gone global. There are programs tailored for use in schools in the UK, for young offenders in New York City, for US Marines awaiting deployment, for firefighters in Florida and taxi drivers in Iran, to name just a few.

But, in intellectually conservative circles, announcing that you meditate is still likely to be met with a snort of derision. Claims in the past about the efficacy of meditation have been tainted with a certain amount of woo-woo New Age mumbo jumbo. In several countries around the world, people still remember elections during the 1990s when candidates for the Natural Law Party advocated transcendental meditation as the cure for all the world's ills. The party declared that its "systematic and scientifically tested" program would involve thousands of meditators creating "coherence in national consciousness" to reduce stress and negativity in society through the power of levitation. I remember watching the party's surreal 1994 European election broadcast in the United Kingdom, which showed cross-legged young men bouncing across mattresses on their bottoms. We were informed that a group of these "yogic flyers" had already reduced the crime rate in Merseyside by 60 percent over the previous seven years.[13]

Against this backdrop, scientists looking into the potential clinical benefits of mindfulness meditation have had to work hard over the past few decades to be taken seriously. Several researchers have told me that when they started out in the field, it was considered career suicide to admit to your peers that you were investigating meditation. This has all changed now. Some of the world's most respected clinical psychologists and neuroscientists are now involved, and their papers are published in mainstream journals such as *Nature, Proceedings of the National Academy of Sciences*, and *The Lancet*. The credibility of the field has been enhanced enormously through the use of new brain-scanning technologies such as fMRI (functional magnetic resonance imaging), which has shown in study after study that meditation produces discernible changes in brain activity.[14]

Another remarkable development has been recent studies of the brains of Buddhist contemplatives who have clocked decades of meditation experience in various monastic traditions. This research has largely been inspired by formal discussions since the 1980s between scientists and the Dalai Lama. One of the neuroscientists most closely involved in this work is Richard Davidson, from the University of Wisconsin, who says we still have much to learn from contemplatives. "This research has underscored the potential value of these traditions for cultivating healthier habits of mind," he told me. "Mental practice can lead to fundamental changes in the brain to support these new habits." He believes that the brain's innate "plasticity"—its capacity for rewiring itself as we learn from experiences and develop new skills—can be harnessed to promote well-being. According to this view, happiness is a skill that, like any other, can be developed through diligent practice.[15]

All the same, there remains a certain wariness of meditation. One common misunderstanding, which inspired my cynical joke that morning at the monastery, is that it transforms people into creatures who have had all their desires, ambitions, and personality excised—zombies, if you will. When I played back the recording of my interview with Ajahn Amaro, I was relieved to discover that it was he who first brought up the subject. I had suggested to him that Buddhism, with its emphasis on cultivating "selflessness," went against the grain of Western culture, with its emphasis on endless striving for self-advancement. It's what gets us out of bed in the mornings and pays our bills. He disagreed. "People think that in Buddhist practice you're meant to be free from desire and so then we shouldn't want anything. They take it to mean that we're supposed to be totally passive, or endeavoring to be a kind of zombie that isn't doing anything. It's a radical misunderstanding, because a) work does not mean suffering, and b) peace does not mean inactivity. When we think 'I want to be peaceful' we think of zoning out at the beach, but you can be completely at peace and working hard at the same time. They are not antithetical to each other."

If anything, this book will argue that the evidence from neuroscience suggests that meditation can make people *less* zombie-like, by giving them more control over their thoughts, emotions, and behavior. *Siddhārtha's Brain* is about the science of mindfulness and the quest for spiritual enlightenment—or, to express the same thing in less loaded terms, the search for optimum psychological well-being. *Enlightenment* has distinct religious overtones, though what Buddhists mean by the word is simply the full realization of the way things truly are—free of any kind of delusion. This is not so different from what scientists are trying to achieve when they investigate the chemistry, physics, and biology of our world. But what of that other slippery word, *spiritual*? As I have looked deeper into mindfulness and Buddhism, the dividing line between the spiritual guidance provided by teachers such as Ajahn Amaro and the mindfulness courses provided by mental health practitioners has started to look less and less clear-cut. In the past decade, thousands of studies have been published that tested the efficacy of secular forms of mindfulness meditation for treating drug addiction, depression, anxiety, and many other afflictions of the mind. Whether you believe this approach is seeking to improve people's "spiritual health" or their "mental well-being" is a matter of perspective. Your choice of words will depend on whether mindfulness training is delivered in a monastery or a clinic. Ajahn Amaro, in common with many other Buddhist teachers, sees himself as much as a mental health counselor as a spiritual adviser. Every day people share their anxieties, their problems, and hang-ups with him. He listens and offers advice about possible courses of action. When it comes down to it, there's not that much difference between his role and that of a secular expert in mindfulness therapy.

More important, how sound is the clinical evidence for the efficacy of mindfulness? New fields of research are often characterized by great enthusiasm among practitioners but also a lack of experimental rigor. Have the benefits of mindfulness been oversold? It wouldn't be the first time a new treatment for mental illness has been hyped by the media and those involved in its development. In 2004, I wrote a feature for *New Scientist* magazine about a class of antidepressants called SSRIs, including Prozac (fluoxetine) and Paxil (paroxetine), which for the past decade had been marketed as wonder drugs.[16] They were said to make you "better than well," with few adverse effects. The popular myth arose that if you took these drugs, you would feel wonderful all the time. When I wrote the feature, that picture was starting to look increasingly flawed, with research coming to light suggesting that the drugs weren't nearly as effective as had been claimed and that they had serious side effects. There followed definitive studies showing that, at best, SSRIs are reasonably effective in mild to moderate depression and at worst they are no use at all.[17–19]

Will clinical applications of mindfulness live up to their early promise, or will they turn out to have been similarly hyped? Is the bubble of enthusiasm surrounding the young scientific field of mindfulness about to burst? In common with many new treatments, the preliminary investigations into mindfulness had some weaknesses, but recent research has been much more rigorous and many analyses have now been published that pool the results of several studies, involving thousands of people in total. The evidence that mindfulness therapy can prevent a relapse in patients with major depression is now very strong, for example. Clinical research into the potential worth of mindfulness programs for treating insomnia, post-traumatic stress disorder, bipolar disorder, psychosis, and many other conditions remains in its infancy, but there is good evidence for its efficacy in anxiety disorders, chronic pain, and drug addiction. Its ability to enhance cognitive performance, such as improving memory and raising IQ, is less certain because not enough high-quality research has been conducted in these areas to date, though there is solid evidence that it can sharpen attention and improve emotional regulation.

What is certain is that unlike popping a pill, mindfulness is no quick and easy fix. To achieve lasting benefits almost certainly requires dedicated practice that continues beyond the standard eight weeks of a training course. Mindfulness is a way of *being* from moment to moment rather than an end in itself, and Buddhists envisage it as just one—albeit essential—element of a much broader program for promoting happiness and contentment. For example, they believe that spiritual enlightenment is impossible without compassion, both for oneself and others, and ethical behavior. One of my aims in writing this book is to bring these teachings to a wider audience and investigate how well they withstand scientific scrutiny.

If mindfulness works as billed, the question then becomes, why? What has gone so wrong during the evolution of the human brain that it needs to be fixed by meditation? Curiously, no one I spoke to during my research for this book had given much thought to this question. So, in *Siddhārtha's Brain* I propose a possible solution to this puzzle based on the available evidence from anthropology, neuroscience, and genetics. There are those who dismiss any speculation about the evolution of mental and psychological traits as belonging in the realm of ideas rather than science. But the human brain and by extension the mind are as much a product of evolution as the eye or the kidney, so it seems odd not to attempt, using all the tools at our disposal, to explain how it developed its strange quirks. If we can then discover exactly how meditation corrects these evolutionary glitches, if indeed it does, we will then have discovered the scientific basis of enlightenment.

Buddhism provides a mental toolkit for improving psychological well-being that was developed in the fifth century BCE, but neuroscientists and psychologists are only just beginning to investigate its potential for changing the brain and behavior. No "longitudinal" study has been published that follows the progress of people in the months, years, and decades after they start to meditate regularly. Suppose, for example, one were to track the changes in the brain of a young adult embarking on such a program, from absolute beginner through to an experienced, even enlightened state many years later? What might that tell us about the potential for fine-tuning the human mind for optimal mental health and happiness? Drawing upon modern scientific evidence, this book travels back in time to envisage such a transformation as it occurred in the brain of an apparently ordinary twenty-nine-year-old man named Siddhārtha Gautama (Siddhattha Gotama in Pāli), who started out on this spiritual journey some 2,500 years ago. He would go on to revolutionize the way his contemporaries viewed themselves and would do more than anyone else to bring the benefits of meditation to our long-suffering species. *Siddhārtha's Brain* presents reconstructions of some of the key moments in this man's life, based on the accounts in Buddhist scriptures.

"We are all mentally ill," said the abbot of Amaravati Monastery with a smile. It was an extraordinary declaration, but I knew exactly what he meant. "That's why we're here," I replied.

CHAPTER ONE

A FOOL'S PARADISE

Our life is shaped by our mind; we become what we think.
Suffering follows an evil thought as the wheels of a cart
follow the oxen that draw it.

—The Dhammapada (translated by
Eknath Easwaran), verse 1

Picture a lush grove on a still, warm evening in late spring. The throb of cicadas and the chatter of a river as it carves a path through the forested landscape are the only sounds. At the center of the grove stands an old fig tree with a broad trunk and fresh green, heart-shaped leaves with long, tapering tips. And, sitting cross-legged beneath the tree, almost hidden in the shadow cast by the setting sun, you might just make out the slight figure of a man wrapped in filthy rags. Look closer and you can't help but notice his sunken eyes, the dark caverns of his cheeks, and how loosely the rags hang from his bony shoulders, though he sits bolt upright—as solid and unwavering as the ancient tree.

Our story begins by the sandy banks of the Nerañjarā River near the village of Uruvelā, in northern India. The birth of Christ lies more than four hundred years in the future, and the foundations of science and philosophy are still being laid by the great thinkers of ancient Greece. The emaciated Indian sitting unmoving in the darkness beneath the tree is Siddhārtha Gautama, a homeless man in his midthirties. Just a few minutes before we arrived he was finishing a meal of rice cooked in coconut milk, scraping the last grains from the bowl. It was his first square meal in a very long time and may well have saved him from a premature, inglorious death by starvation. Describing his predicament in later life, he would say that after years of brutal self-denial his hair had started to fall out. His limbs looked "like the jointed segments of vine stems or bamboo stems. Because of eating so little my backside became like a camel's hoof." His ribs jutted from his chest like "the crazy rafters of an old, roofless barn," his eyes had sunk so deep into their sockets they were "like the gleam of water deep in a well."[1]

His father, Suddhodana—a wealthy and influential man who was the elected chief or "king" of the Shakya clan in their remote northern republic in the foothills of the Himalayas—would have been horrified to see him in this state. Six years previously, Prince Siddhārtha was living in great comfort in the grand family home at Kapilavatthu (Kapilavastu), the republic's capital, about 230 miles northwest of Uruvelā, near the border between what is now southern Nepal and the Indian state of Uttar Pradesh. His family were members of the governing warrior class, the *kshatriya*. According to legend, when Siddhārtha was a baby, eight brahmin

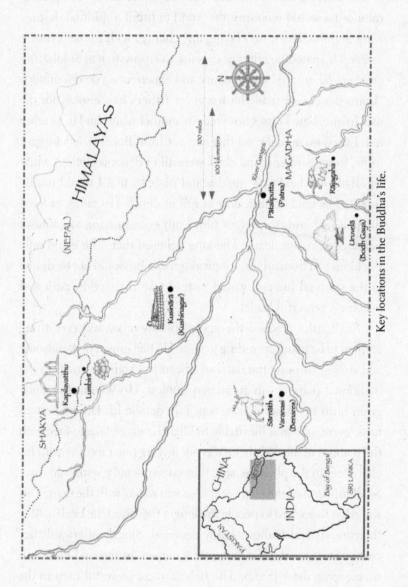

Key locations in the Buddha's life.

priests foretold that he would become either an all-conquering ruler or he would renounce the world to fulfill a spiritual destiny. King Suddhodana wasn't taking any chances with his son's future career. He spared no effort or expense to ensure that as Siddhārtha grew up, he enjoyed every luxury and experienced no discomfort. "Lotus pools were made for me at my father's house solely for my use; in one, blue lotuses flowered, in another white, and in another red. I used no sandalwood that was not from Benares. My turban, tunic, lower garments and cloak were all of Benares cloth. A white sunshade was held over me day and night so that I would not be troubled by cold or heat, dust or grit or dew." [2] His father ordered the palace guards to prevent him from encountering any hint of sickness, aging, or death. The king believed that if he could only shield his son from all life's unpleasantness, he would not be drawn to the spiritual life and would instead take the worldly path and become a powerful leader.

Siddhārtha reached the age of twenty-nine, and everything seemed to be going according to plan. He had grown up handsome and strong, winning the hand of a beautiful young woman in the traditional manner—in an archery contest. His wife had recently given birth to a healthy baby son. But, despite his father's best efforts, eventually and inevitably Siddhārtha came face-to-face with the realities of life. While he was out driving one morning with his charioteer in the pleasure park, they encountered a senile old man. Siddhārtha asked the charioteer what was wrong with the man. This was what happened to people the longer they lived, he explained—their minds and bodies steadily declined. Shortly afterward they came across a sick man and later a corpse. In the end there was no escaping these things. The richest, most powerful man in the world could not hold back sickness, aging, and death. It dawned on

Siddhārtha that sooner or later even the most beautiful and wonderful things in his life—the most sensual pleasures—would fade. Nothing would be perfect, nothing permanent. Everything he had come to love was subject to change, death, and decay.

The king probably noticed a change in his son's demeanor. He seemed distracted, depressed. To cheer him up, that evening Suddhodana sent dancers and musicians to entertain him. But as Siddhārtha would later recall, when he awoke on his couch in the middle of the night the performers had all fallen asleep. The minstrels had let their instruments slip through their fingers and the dancing girls had flopped exhausted to the floor. They were a pitiful sight, "some drivelling at the mouth, spittle-besprinkled, some grinding their teeth, some snoring, some muttering in their sleep, some gaping, and some with their dress in disorder . . ."[3] The scene filled Siddhārtha with disgust. What had been bright, sensual, and beautiful just hours before had become ugly and debased. So this is what follows in the wake of worldly pleasures, he thought. When he retired to his bedchamber and gazed upon his sleeping wife, desire died in him because all he saw was the old woman she would become. Looking down at his newborn son in his cradle and pondering the boy's future, all Siddhārtha saw was a trap holding them both fast to this futile round of duty, trivial entertainments, pain, disappointment, and death.

Faced with this sudden mental crisis, the solution seemed obvious. He would run away and start a new life free from the fetters of home and family. He would seek a path out of this cycle of suffering. Earlier that day, after encountering the horrors of sickness, aging, and death for the first time while riding in his chariot through the town, another strange being had caught his eye: a man sitting cross-legged on a street corner, apparently untouched by the noise and chaos all about him, radiant and serene. His charioteer told him this was a wandering ascetic, a truth-seeker who lived in the forests and relied on the generosity of others. For Siddhārtha, he was like a heavenly messenger showing him the way. "While still young, a black-haired young man endowed with the blessings of youth in the first stage of life—and while my parents, unwilling, were crying with tears streaming down their faces—I shaved off my hair and beard, put on the ochre robe and went forth from the home life into homelessness."[4] So began his search for the "incomparable peaceful state" of spiritual enlightenment.

Despite the passage of two and a half thousand years, we can empathize with the spoiled young Siddhārtha's plight. Like him, many of us have been raised in a fool's paradise. Most people in the developed world have food in abundance; entertainment and pleasant distractions are just a short journey or finger tap away; drugs and surgery give us the illusion that we can defeat disease and aging (though in reality they simply delay and prolong old age). Until comparatively recently in human history, encounters with death were commonplace, but young people nowadays find it almost impossible to imagine that one day they too will die. Like Siddhārtha, many grow up without ever having seen a dead body with their own eyes. Death is a staple of movies, television dramas, and the news, but our own demise is a profoundly taboo topic of conversation. Perhaps we subconsciously believe that if we don't talk about it, we can somehow cheat it. For much the same reason, we are loath to talk about debilitating or fatal illnesses such as cancer. These delusions can't last, of course, but maybe they are worth sustaining for as long as possible if they allow us to live happy, fulfilled lives while we still have our health. If only it were that simple. In advanced economies, many of us have enjoyed steadily rising standards of living since the 1950s and taken advantage of increasingly sophisticated social-support and health-care systems, and yet self-reported levels of life satisfaction have scarcely budged in more than half a century. We are in the grip of what epidemiologists call the "happiness paradox."[5]

Are we doing something wrong? Research suggests that, ahead of physical health, employment, and freedom from poverty, mental health is the most important determinant of individual happiness in developed countries. Unfortunately, we appear to be failing badly on this score.[6] The World Health Organization (WHO) estimates that, globally, 450 million people suffer from a mental health or behavioral disorder (350 million of them are adults suffering from clinical depression), which makes psychiatric illness one of the leading causes of ill health and disability. Worldwide, nearly a million people commit suicide every year.[7] Startling rates of psychiatric illness are seen even in the richest countries. In the UK, one in four people will face some kind of mental health crisis in the course of a year, with chronic anxiety and depression being the most common afflictions. Almost 6 percent of Britons over the age of sixteen report that they have attempted suicide at some time in their lives.[8] Before too long, mental illness will be putting a greater strain on the health services of rich countries than any other disease, with the WHO predicting that depression will be the biggest "disease burden" in high-income countries by 2030.[9] The global cost of mental disorders in lost economic output is predicted to be $16 trillion over the next twenty years.[10]

In 2015 I attended a conference in London entitled The Global Crisis of Depression. Former UN secretary general Kofi Annan opened the proceedings with these words: "Let's be honest, sometimes the title of a conference can overstate challenges in an understandable attempt to focus attention on a neglected issue. But that is not the case today. Calling the challenge of depression a global crisis is no exaggeration at all." In chapter 6 of this book, "Golden Slippers," I explain the role of a psychotherapy inspired by Buddhist contemplative practices, known as mindfulness-based cognitive therapy (MBCT), in efforts to tackle this daunting challenge.

Affluence doesn't shield us from unhappiness, though it certainly helps. There is a clear association between poverty and mental illness. The legend of Siddhārtha's early life looks almost like a thought experiment conducted by his early followers as they ornamented and passed down the story to future generations: Suppose a human had everything he or she could possibly want—physical health, good food in abundance, riches, comfort, sensual pleasures, status, a family, security—would this be sufficient to bring lasting happiness? Their conclusion was that it would not: the human psyche is inherently flawed, preventing lasting satisfaction even when circumstances seem ideal.

This was a shocking discovery. Where did it all go wrong for the human mind? It is easy to fall into the trap of thinking that evolution proceeds like the design process of a series of increasingly complex motor cars—a neat, ordered progression from the equivalent of a Ford Model T to the power and sophistication of a Formula One racing car—when in fact it has been a messy and imperfect process. We're still here, our species has thrived like no other, but there have been plenty of mishaps along the way. Evolutionary breakthroughs have their downsides. To give just a few examples relevant to human health, the vertebrate immune system evolved over the course of billions of years to protect the body from invasion by pathogens, but it can also turn against our own tissues to cause a wide range of common autoimmune diseases, including rheumatoid arthritis, multiple sclerosis, and type 1 diabetes. Cells replicate to renew tissue and repair damage, but they can also divide uncontrollably to cause cancer. People who have a single copy of a particular gene can tolerate infection by the *Plasmodium* parasite that causes malaria,[11] which has been rife in sub-Saharan Africa ever since the development of agriculture thousands of years ago. But having two copies causes the excruciating and potentially fatal condition known as sickle cell anemia.

Natural selection, rather than bringing us closer and closer to a state of godlike perfection, is in reality a series of awkward compromises. Adaptations may have a net benefit, but they bring disadvantages in their wake. So it has been with the evolution of the human mind. There can be no doubt that our brain, which Isaac Asimov[12] once described as "the most magnificently organized lump of matter in the known universe," is a wonderful product of natural selection, with language and creativity among its unique adaptations, and yet the statistics for mental illness suggest that it has not been crafted particularly well to provide psychological stability and lasting happiness. Natural selection weeds out genes from populations that harm an individual's ability to thrive and reproduce, and at first glance common mental illnesses such as addiction, anxiety, and depression don't fit this universal law. Even though they have a strong genetic component and tend to reduce the number of children that patients will have compared with people who don't suffer from these conditions, they have nonetheless remained widespread in populations all around the world.[13] This suggests that the very same genes that make some people more susceptible to common mental illnesses have also played a vital role in ensuring our survival as a species. They have conferred disadvantages as well as advantages.

The exact nature of these trade-offs has yet to be established, but you don't have to look very far to find examples of the costs and benefits provided by the human central nervous system. We have built-in biological drives such as hunger, thirst, and sexual desire that are indispensable for the perpetuation of our genes. The neurotransmitters of the brain's reward system ensure that we fuel our bodies and reproduce. But this is the same system that gives us a kick from eating a huge tub of chocolate chip ice cream at one sitting—or snorting a line of cocaine. And the reward system not only keeps us coming back for more of the pleasurable substance or activity, it can also become less responsive after repeated hits, which means greater doses of the particular drug, food, or behavior are needed to achieve the same effect. Perhaps by pampering Siddhārtha and providing every pleasurable distraction, his father unwittingly brought about the very destiny he was seeking to avoid for his son. In a world of plenty, the drives that help us survive in more challenging environments can become the cause of our downfall, drawing us into a cycle of craving, overindulgence, disappointment, and regret. This is the subject of chapter 7, "Fire Worshippers," which explores addiction and some promising research that suggests that meditation can be used to reduce drug cravings, helping smokers to quit and former drug addicts to stay clean and sober.

I will argue that our many mental weaknesses can be traced back to the hardwired responses that allowed our ancestors to thrive under somewhat different circumstances in our distant evolutionary past. Another example is the fight-or-flight response—the series of physiological changes orchestrated by the central nervous system that prepares our bodies for combat or to run for our lives. This will have meant the difference between life and death for an early human knocked to the ground by a hungry predator, but an alarming sensory stimulus such as a sudden loud noise or being shoved on a busy underground train triggers exactly the same changes in the body. It goes without saying that punching the person who has accidentally shoved you on an overcrowded train is not going to make your life, or theirs, a whole lot happier. Worse still, in the long run, prolonged activation of the fight-or-flight response—otherwise known as chronic stress—is physically and psychologically harmful, putting us at greater risk of heart disease and mental illness.[14, 15] In the next chapter, "Child's Play," I will introduce the relaxation response, which is the body's natural counterbalance to the fight-or-flight response. It is now well established that meditation is very effective at evoking this physiological response, helping people to cope with stressful situations by giving them the ability to restore their bodies at will to a less emotionally aroused state. I will provide some simple instructions you can use to start practicing this form of meditation yourself. There are several other meditations scattered throughout the book to give you a flavor of other common mindfulness techniques.

Addiction and chronic stress are among the more obvious "design flaws" that have crept into humans' mental blueprint in the course of our evolution, flaws that have become apparent relatively recently in our species' history. Fortunately, the brain has another inbuilt mechanism that over time can help turn down overpowering emotions, such as anger and fear, which is the subject of chapter 8, "A Drunk Elephant." But the problems for the brain don't end with our primitive drives, emotions, and defense mechanisms. Some abilities that set us apart from all the other primates—including language, creativity, and the capacity to live in very large cooperative groups—also have their downsides, as I will explain in chapter 9, "The Fall." These are talents that have been added to our mental toolbox in the course of our evolution into intelligent, highly sociable apes. Our brains have evolved a capacity known as "theory of mind" that allows us not only to recognize ourselves as individuals separate from our fellows, but also to step metaphorically into others' shoes, seeing the world from their perspective and attributing to them beliefs, thoughts, and desires. This faculty allows us to predict how others will behave in a particular situation, or work out why they have said or done something. It gives us the ability to empathize with their feelings, but it also affords us the capacity to deceive them. And, if an individual's theory of mind is poorly tuned, he or she may misinterpret others' motivation or intentions, which can lead to delusions and paranoia.[16]

We are also considerably better than our closest primate cousins at mental time travel. In the theater of our minds, we can relive personal experiences such as conversations, food we have tasted, and music we have heard. We can range forward, speculating about future events and working out what we will do, what we will say, anticipating how we will react in particular circumstances. It's the stuff of thought, essential for reasoning, planning, and learning from experience, and it is the brain's default mode when it is not performing a particular external task. Unoccupied, the mind runs here and there of its own accord, pursuing trains of thought like a frisky dog let off the leash in a park. Even when we're trying to concentrate on something important, such as writing an email, talking to a friend or colleague on the phone—or, dare I say, reading a book—our attention is constantly straying.

A wandering mind is certainly not a very effective mind, but could it be making us unhappy? By their very nature, thoughts are subjective and as fleeting as gusts of wind, but psychologists have done their best to get to grips with them using a technique called "experience sampling," in which people are asked to report in a journal they carry around, or to a researcher on the end of a phone line, what they're thinking and feeling at predetermined times during the day. But this kind of study is costly to administer, and not very convenient or naturalistic for the participants, so sample sizes are small and the results are unreliable. Psychologists at Harvard have found a typically twenty-first-century solution to this problem: they created an app. In the past few years, more than fifteen thousand people have downloaded their iPhone app, called Track Your Happiness, which at intervals during the day interrupts what they are doing to ask them questions such as *How are you feeling right now?*; *What are you doing right now?*; and *Are you currently thinking about something other than what you're doing?*

Based on the responses of 2,250 adults, the researchers concluded that, overall, people's minds wander from what they are actually doing an amazing 47 percent of the time, and for at least 30 percent of the time that they spend on any activity . . . apart from making love (10 percent).[17] As John Lennon sang, "Life is what happens to you while you're busy making other plans."[18] On the whole, the type of activity had only a modest effect on whether or not the volunteers' minds wandered and had no impact at all on whether the distracting thoughts were pleasant or unpleasant. Crucially, people reported feeling less happy when their minds were wandering compared with when they were not, regardless of what they happened to be doing. So, even if they were doing their least favorite thing, such as housework or commuting, a wandering mind made them feel even less happy. Statistical analysis of the data suggested that a wandering mind was the *cause* and not merely the consequence of unhappiness. Remarkably, what people were *thinking* seemed to be a better predictor of their happiness than what they were *doing*.

Psychologists were not the first to notice this phenomenon. In the Dhammapada (The Path of Truth), a collection of sayings attributed to Siddhārtha, the first pair of verses neatly sums up this universal rule:[19]

> *Our life is shaped by our mind; we become what we think. Suffering follows an evil thought as the wheels of a cart follow the oxen that draw it.*
>
> *Our life is shaped by our mind; we become what we think. Joy follows a pure thought like a shadow that never leaves.*

Siddhārtha would come to believe he had found the antidote to suffering. He thought he could fix the flaws in the human psyche. In the fifth century BCE, people were not even dimly aware that the mind is a product of the electrical activity of the brain, itself the product of billions of years of evolution. It seems he didn't need to know such things to develop his model of the human mind. His philosophy, lived out in his own quest for enlightenment, was that you should try different practices and see for yourself whether or not they work. If someone's teaching doesn't add up, if it doesn't ease suffering or makes it worse, you abandon it. Skepticism is encouraged. He once said, "Don't go by reports, by legends, by traditions, by scripture, by logical conjecture, by inference, by analogies, by agreement through pondering views, by probability, or by the thought, 'This contemplative is our teacher.' When you know for yourselves that, 'These qualities are unskillful; these qualities are blameworthy; these qualities are criticised by the wise; these qualities, when adopted and carried out, lead to harm and to suffering,' then you should abandon them."[20]

In other words, *nullius in verba*—"take nobody's word for it." This is the motto of the Royal Society, founded in London in the seventeenth century to promote a new kind of philosophy that rejected received wisdom and instead sought knowledge through observation and experiment. Back then it was called "natural philosophy," nowadays we call it science. Of course, the kind of observation that Siddhārtha advocated was *self*-observation. This may be invaluable for gaining insights into one's own emotions, behavior, and motivation, but what of the insights passed on to us by others? Is it rational to trust the insights of a few people, no matter how venerable, into their own minds and apply the lessons to everybody? Fortunately, we no longer need to take this leap of faith: science has provided us with objective tools such as clinical trials and technologies such as genome mapping and magnetic resonance imaging that can be used to test particular claims with unprecedented rigor. We can probe scientifically not only *whether* meditation and other elements of Buddhist practice have tangible benefits, but also *how* they might operate in the brain to influence behavior and well-being.

Buddhism is perhaps the most peaceable and down-to-earth of all world religions. It is not a belief system, and to practice it, one is not required to recite a creed or commune with gods, angels, or the souls of the departed, but rather to investigate the ways of one's mind. While it does not have a clean record by any means, it has a history of accommodation with other religions, most notably in its country of origin, India. One of my prime motivations for writing this book was to explore the scientific credibility of its psychology. Buddhists assert, for example, that we can minimize suffering and maximize well-being through regular meditation and adherence to a strict code of behavior and thought. They believe these practices change the brain for the better. Neuroscientists have known for a long time that the brain is "plastic," with new nerve cells and connections being formed and destroyed throughout our lives in response to what we experience through our senses. Learning involves the creation of new synapses—the electrical contacts that allow nerve cells to communicate—which is the basis of memory, the development of new habits and the dissolution of old ones, and the learning of new skills. Experience drives these changes. So you could say that not only is our life shaped by our mind, our brain is shaped by our life. The objective of Buddhist practices is to harness this process to promote psychological well-being. As the Dhammapada has it, "The farmer channels water to his land, the fletcher whittles his arrows, and the carpenter turns his wood. So the wise direct their mind."[21]

The core practice is known as "mindfulness," which involves making a conscious effort to live nonjudgmentally in the present moment, acknowledging thoughts, feelings, and sensations as they arise and accepting them just as they are. This practice is understood to help one deal with psychological challenges more objectively, rather than with automatic responses based purely on emotions, fears, and preconceptions. Over the past few years there has been a surge of popular interest in the secular form of this ancient mental discipline, with training courses springing up all over the world and becoming available online and via apps. Scientific journals have published studies by psychologists and therapists suggesting that this deceptively simple technique can not only help treat pain, anxiety, depression, and drug addiction but also improve everyday concentration and performance. There have even been hints that it could slow the aging process and keep dementia at bay, a possibility I explore in chapter 11, "Mind Mirrors."

Many claims about mindfulness have undoubtedly been overstated; most of the early studies were small and imperfectly designed. But there is increasingly solid evidence for its clinical benefits. For example, analyses of the most thorough research to date suggest that mindfulness is as effective as antidepressants for treating mild depression[22] and *more* effective than the drugs for preventing relapse in people with recurrent major depression who experienced severe abuse during their childhood.[23, 24] There is also good evidence that it can counteract anxiety and stress, and help reduce the severity of chronic pain.[25] When American scientist Jon Kabat-Zinn developed the world's first secular mindfulness course in 1979, the first to try it out were patients who had been experiencing severe pain for several years that had never been adequately controlled by painkillers or surgery. I bumped into Kabat-Zinn in 2014, in a hotel elevator at a mindfulness conference in Boston, and he generously agreed to an interview. In chapter 4, "The Second Dart," I describe how his own experiences practicing Zen Buddhism as a student inspired him to adapt some of these ancient practices to help patients cope with chronic pain, anxiety, and stress.

Functional magnetic resonance imaging (fMRI) scans indicate that as little as eight weeks practicing mindfulness meditation can bring about observable changes in a beginner's brain. An ongoing collaboration between scientists and the Tibetan spiritual leader, the Dalai Lama, is providing evidence that thousands of hours of meditation by Buddhist monks and nuns over many years has wrought a much more dramatic transformation in their brains. This becomes apparent when they are compared with the brains of people who don't meditate. The question remains whether those differences have arisen as a result of meditation or whether they were there all the time. Perhaps people with this brain type are more likely to choose a life of quiet contemplation? To differentiate between these alternatives, ideally one would need to scan the brains of people before they begin their monastic life and then repeatedly over the years and decades to track any changes. Unfortunately, longitudinal studies of this kind are extremely rare due to their cost and the challenges involved in organizing and administering them.

Drawing upon the best published research and interviews with scientists, I will investigate exactly how the practice of mindfulness might bring about changes in dedicated meditation practitioners like Siddhārtha and how they would have influenced his behavior and well-being. In chapter 5, "The Man Who Disappeared," I will discuss the scientific evidence for what is perhaps his most revolutionary teaching—which remains deeply controversial and counterintuitive to this day—that there is no such thing as an unchanging, distinctive "Self" living in our heads. Despite losing our souls, we still have left to us the mysterious gifts of consciousness and the related ability to "think about thinking," which are the subjects of chapter 10, "Wonderful and Marvelous."

Regardless of claims about the nonexistence of an unchanging Self, at the core of Buddhism is a prescription for improving one's well-being in the here and now. Uniquely among world religions, it doesn't impose a creed on its followers. It doesn't demand that they believe in the supernatural. That is not to say that as individuals, Buddhists aren't superstitious—far from it. Across Asia there are many who still believe in spirits, ghosts, and gods. Most trust that after death they will be reborn in another body and that certain acts, such as giving offerings of food to monks or making donations to their local temple, earn them "merit" that can help bring about a favorable rebirth in the next life. In the final chapter, "The Deathless Realm," I will address these beliefs and propose an updated version of *kamma* that offers an optimistic vision for the future of our species. (You may be more familiar with the Sanskrit spellings of Buddhist terms, such as *Dharma, karma,* and *nirvana,* than you are with the Pāli spellings, *Dhamma, kamma,* and *nibbāna,* favored by Theravada Buddhists, including the Thai Forest tradition.)

Superstitions and ritual practices have much deeper roots in India than even Buddhism. But at its heart is a program for minimizing suffering and promoting well-being devised two and a half thousand years ago by a vagrant called Siddhārtha Gautama, drawing upon little more than his own experience, close observations of human life, and an unflinching exploration of his own mind. Crucially, other contemplatives and philosophers have reached similar conclusions through their own investigations. I give a flavor of this convergence in chapter 3, "The Cloud of Unknowing." Above all else, Siddhārtha believed that to achieve enlightenment, one must see the world as it really is, in all its fearsome impermanence, stripped bare of any kind of delusion. He taught that one must find out the truth for oneself, not taking anybody else's word for it. Surely this is the ideal mind-set of a scientist? I feel confident that Siddhārtha would have welcomed the light that modern science is now shining on his formula for enlightenment. And I suspect that, like many twenty-first-century monastics, he would have been perfectly happy to help out neuroscientists with their research.

Before we continue, though, a little historical perspective is in order. In the fifth century BCE, when Siddhārtha abandoned his life of luxury and embarked on a quest to discover the antidote to human suffering, he was joining many thousands of others on the Gangetic plain who had renounced society and were on that same spiritual journey. There were already bands of wandering ascetics who followed inspirational leaders, and others who lived alone in the forest, passing their time in contemplation. They were all part of a popular movement that had been rebelling against the religious conservatism of their times. Seven centuries earlier, Aryan invaders from the north had established a religion-based society in India that had become ossified into a hierarchy of hereditary castes. At the top were brahmins, priests who told people how they should live and maintained a close symbiotic relationship with regional chieftains or kings; then came the warrior kshatriya class to which Siddhārtha belonged, responsible for government and defense; then the *vaiśya*, or vaishya, who tended the land; and finally the *shudra*, or sudra—artisans and laborers buried at the bottom of the heap. The brahmins, under the influence of the hallucinogenic drink soma, were believed to be channels of the universal law that governed the lives of gods and men. They preserved this lore orally in texts known as the Vedas, which brahmin fathers passed to their brahmin sons in Vedic Sanskrit—a language that no one else understood. They were the keepers of the sacred fires, which they tended in shrines and never allowed to go out. They chanted the ritual verses of the Vedas and performed the blood sacrifices that they believed kept the world in existence.[26]

From around the sixth century BCE, however, the old hierarchical society had been fracturing. New Iron Age technologies had increased the productivity of farming and allowed forests to be cleared to plant more crops, creating surpluses of food for trade and allowing more and more people to leave the land and move to the growing cities, which became the centers of production for manufactured goods such as textiles. The luxurious cloth that Siddhārtha wore in his youth was from one of those cities, Varanasi (also known as Benares). To facilitate all this commerce, a new class of merchants, bankers, and businessmen had sprung up— people who were no longer tied by the hereditary bonds of caste, king, and priesthood. The wealth and urban lifestyle they created around them brought more time for thinking, talking, speculating about the meaning of life, and even questioning the authority of the brahmins on spiritual matters. The merchants established trade routes that not only made possible the long-distance transport of surplus food and luxury goods such as spices, jewelry, and fabrics, but also ideas. Even in his father's palace in remote Kapilavatthu, Siddhārtha must have felt the pull of these radical ideas.

For many it must have been like waking from a long, deep sleep, opening their eyes, and trying to focus on their surroundings, only to find themselves in an unfamiliar place. Being freed from the old certainties taught by the brahmins was liberating, but it was also bewildering. Nothing made sense anymore. Life tasted bitter, and everywhere one looked, there was a nagging sense of unsatisfactoriness. Making it all worse, human beings believed themselves to be caught up in a never-ending cycle of birth, death, and rebirth, condemned to undergo the tortures of illness, aging, and annihilation again and again. Imagine life in an era before antibiotics, vaccination, and pain relief, and then imagine the prospect of having to face this ordeal of disease, pain, and death repeatedly ad infinitum. This cycle of rebirth was called *saṃsāra*. There was some hope of improving your lot, because according to the law of kamma (karma), if your deeds in this life were good and wholesome, in the next you could be reborn as a wealthier person higher up the social hierarchy, or even into the realms of the gods. But if your life was ruled by desire, cruelty, and dishonesty, you would be reborn into a lower caste or, worse still, as an animal.

The "renouncers" were wandering folk who, like Siddhārtha, had willingly made themselves homeless and were looking for a way to escape this cycle and attain an existence free of suffering: they sought enlightenment, or nibbāna (nirvana). They thought they could achieve this through brute exertion, giving up any kind of comfort or pleasure in the hope of making progress toward their spiritual goal. In their eyes, the way people made a living in the new cities was inherently flawed because it was driven by desire and ambition. These were the attributes that made the world of commerce go round, but they also turned the wheel of suffering. Above all else, the renouncers were looking for truth and meaning at a time when these things seemed to have been lost in the head-long rush for material and social advancement. There were leaders among them who had their own prescription for enlightenment, and each had a band of followers earnestly trying out the teaching to see where it would lead.

So, when Siddhārtha left his father's house, he wandered the kingdoms and republics on the plains of the Ganges looking for a suitable teacher, eventually joining the followers of Ālāra Kālāma, a Yogi who taught that nature was ephemeral and that to end suffering one must rise above it to discover Atman, the eternal, unchanging Self that was indistinguishable from the essence of the universe. This core of the person was untouched by the body, with its fickle emotions and primal urges. Yoga in its original form had little to do with health and relaxation: it was about mastering the senses and subduing the egotistical, mundane self and its constant distractions. Only by stripping away your crude nature could you experience the bliss that was the undying Self. Thousands of years before Sigmund Freud would write about the subconscious, Yogis in ancient India had identified the untamed mind as a principal source of suffering.

To free their minds, the followers of Kālāma adhered to a strict moral code: no lying, stealing, harming any other living creature, alcohol, or sex. They learned to endure hunger, thirst, heat, and cold without complaint. Every urge that anchored them to their animal nature was ruthlessly suppressed. Finally, they attempted to sever the link between their mind and their body once and for all by sitting motionless for hours on end like something dead, deliberately slowing or even stopping their breathing. These disciplines were said to lead to an altered state of consciousness called the "sphere of nothingness," which Kālāma claimed to be Atman. But even though Siddhārtha became a proficient yogin, spending several years taming his senses and honing his contemplative skills, he did not attain nibbāna. Deep meditation liberated his mind, but when he rose to the surface of ordinary consciousness he was still the same man with all his animal urges and angst intact. He suffered just as much as ever.

Disillusioned, he latched onto another teacher, Uddaka Rāmaputta. But the same thing happened. He learned this Yogi's techniques and developed the requisite mind-set until he had outshone his new mentor, but he remained unchanged. So he struck out on his own. Soon he was developing a modest reputation as a sage in his own right with five followers. Together they practiced the most extreme forms of asceticism a human can endure. "I took food once a day, once every two days . . . once every seven days, and so on up to once every fortnight," Siddhārtha would later recall. "I was an eater of greens or millet or wild rice or hide-parings or moss or rice bran or rice scum or sesamum flour or grass or cow dung. I lived on forest roots and fruits, I fed on fallen fruits. I clothed myself in hemp, in hemp-mixed cloth, in shrouds, in refuse rags, in tree bark, in antelope hide, in strips of antelope hide, in kusa-grass fabric, in bark fabric, in wood-shavings fabric, in head-hair wool, in animal wool, in owls' wings. I was one who pulled out hair and beard, pursuing the practice of pulling out hair and beard. I was one who stood continuously, rejecting seats. I was one who squatted continuously, devoted to maintaining the squatting position. I was one who used a mattress of spikes."[27]

The objective was not only to torment and mortify the body but also to reject society and its norms. It was as though Siddhārtha no longer wished to be human. "I would make my bed in a charnel ground with the bones of the dead for a pillow. And cowherd boys came up and spat on me, urinated on me, threw dirt at me, and poked sticks into my ears." But no breakthrough came, no enlightenment. "I thought: 'Suppose that I, clenching my teeth and pressing my tongue against the roof of my mouth, were to beat down, constrain, and crush my mind with my awareness.' So, clenching my teeth and pressing my tongue against the roof of my mouth, I beat down, constrained, and crushed my mind with my awareness. Just as a strong man, seizing a weaker man by the head or the throat or the shoulders, would beat him down, constrain, and crush him, in the same way I beat down, constrained, and crushed my mind with my awareness. As I did so, sweat poured from my armpits. And although tireless persistence was aroused in me, and unmuddled mindfulness established, my body was aroused and uncalm because of the painful exertion."

Then, as the Yogis had taught him, he tried to stop his breathing. "As I did so, extreme forces sliced through my head, just as if a strong man were slicing my head open with a sharp sword. . . . Extreme pains arose in my head, just as if a strong man were tightening a turban made of tough leather straps around my head. . . . Extreme forces carved up my stomach cavity, just as if a butcher or his apprentice were to carve up the stomach cavity of an ox. . . . There was an extreme burning in my body, just as if two strong men, grabbing a weaker man by the arms, were to roast and broil him over a pit of hot embers." Surely, he thought, no brahmin or contemplative had ever endured greater pain? But it was all for nothing.

Siddhārtha was now close to death, in the emaciated state in which we found him at the start of this chapter. Despite all he had learned and all he had endured, he felt no closer to his ultimate goal. Six years had passed since he left home. Home . . . In a daze, he remembered his boyhood and an afternoon when he had felt truly at peace.

CHAPTER TWO

CHILD'S PLAY

Do naught with the body but relax;
Shut firm the mouth and silent remain;
Empty your mind and think of naught.
Like a hollow bamboo rest at ease your body.

—Tilopa, "Song of Mahamudra,"
translated by Garma C. C. Chang

The bullocks strain in their harness as the iron blade cuts through the dry compacted soil. Guiding the plow with one hand and wielding a whip in the other, King Suddhodana plods along behind them in the furrow, a look of dogged concentration on his face. To one side, on the unplowed ground, an attendant fusses with a long-handled ceremonial parasol, trying in vain to shade his master's head from the ferocious sun. Next come the palace guards holding aloft pennants emblazoned with the lion insignia of the Shakya clan; then the court musicians beating their drums; and finally the brahmins in their colorful finery, chanting incantations as they cast rice into the broken earth where the little creatures exposed by the plow writhe and scurry.

Hundreds of the king's subjects have turned out to watch him plow. They crowd the margins of the field in respectful silence. Yet the royal enclosure, on rising ground at one corner, is all but empty. Only half a dozen loyal courtiers have braved the midday sun to observe the ceremony. Twenty yards from the field, on a hillock at the far end of the tightly guarded enclosure—alone and apparently forgotten—a small boy of seven or eight sits cross-legged on a rug in the generous shade of a tree, a toy bow and arrow at his side. This morning Siddhārtha's nursemaids dressed him in his finest green and gold silk outfit for the annual festival, but now they have abandoned him for a closer view.

The crack of his father's whip, the low murmur of the brahmins' voices, and the beat of drums grow louder as the plow nears the royal enclosure, carving a crooked line in the dirt. The boy observes everything intently. A commotion ensues when the king struggles to negotiate the corner of the field, bringing the procession to an abrupt halt—priests, musicians, palace guards, and servants colliding—but with some help from his men-at-arms, the king masters the cumbersome plow, the two bullocks are whipped and cajoled into position, and the ceremonial procession starts up once more, receding toward the far corner of the field. The drumbeat and chants fade in the boy's ears. A breath of wind stirs the branches above his head. His eyes close.

According to Buddhist lore, one day while his father officiated at a spring plowing ceremony, the eight-year-old Siddhārtha was left alone in the cool shade of a rose apple tree, where he fell into a state of profound calm.[1] Twenty-seven years later, as a starving ascetic almost at death's door after six years of relentlessly subduing his senses and punishing his body, he would remember this happy childhood experience and wonder whether it could be the first step on the path to enlightenment he had been seeking for so long.

The boy prince wasn't the first person to discover this particular mental technique for relaxing the body and mind, and he surely won't be the last. Throughout recorded human history and probably much further back in time, humans have stumbled upon the portal that leads into this tranquil state of being. Christians, Hindus, Muslims, and Jews have described the experience in purely religious terms, though there is nothing intrinsically mystical about it. Some of the earliest Western research into the phenomenon was inspired by its potential medical benefits, but the experimental subjects were monkeys rather than people. In the late 1960s, Herbert Benson, a brilliant young cardiologist recently graduated from Harvard Medical School, was intrigued by the way his patients' blood pressure was often higher when he measured it in his consulting room than when they recorded it for themselves at home or if it was monitored automatically throughout the day by a portable device. None of his colleagues seemed bothered to investigate this "white coat hypertension," as it had become known, but Benson had an inkling what might be going on. He had a hunch that what was raising their blood pressure was his patients' heightened anxiety in the presence of their doctor and in the austere surroundings of a medical clinic. To readers in the twenty-first century, the idea that the mind has a powerful influence over the body and can change the course of an illness won't seem in the least surprising, but at the time the concept of "psychosomatic" effects like this was highly controversial. Even today, to drop into conversation that someone's illness is psychosomatic would be to imply that it is "all in his mind," as if that somehow made him less worthy of our sympathy. So, when Benson decided to go back to the labs at Harvard Medical School to investigate his hunch about stress and hypertension—using squirrel monkeys—his colleagues thought he had taken leave of his senses.

Benson set about training the monkeys to lower their own blood pressure.[2] The technique is called biofeedback and had been pioneered just a few years previously by scientists, such as Neal Miller at Yale University, who trained people and laboratory animals to change particular aspects of their physiology including their heart rate and the electrical activity of their brains. First, Benson used green or red lights to signal to the monkey when its blood pressure was going up or down. The lights were accompanied either by a punishment—a mild electric shock—or a reward of food. Soon Benson found he could raise or lower the animals' blood pressure simply by switching between the two colored lights, without having to deliver a punishment or reward. Their bodies had learned to associate the lights with unpleasant or pleasant stimuli. Like the patients in his clinic, a particular feature of the monkeys' environment was now directly influencing their physiology. This was standard classical conditioning—the very same learning process discovered by Pavlov with the help of his famous drooling dogs at the turn of the twentieth century. But what happened next was much more impressive. The monkeys seemed to learn how to lower their blood pressure *at will* in order to receive food, even in the absence of the colored lights. They were now controlling an aspect of their physiology that was previously involuntary and governed solely by their visceral, "autonomic" nervous system. It was at this point, in the late 1960s, that Benson's research attracted the attention of folk in the transcendental meditation movement.

In 2014, I was amazed to discover that Benson, at seventy-nine years old, was still working, five decades after these experiments took place, at an age when most of us would be content with a little light gardening or doing a crossword puzzle. He seemed perfectly happy to talk to a stranger on the far side of the Atlantic about his early years as a cardiologist and meditation researcher. When I phoned him at his office in Boston, he was in top form, eager to explain his life's work. He recalled perfectly the day in 1968 when young followers of the Indian mystic Maharishi Mahesh Yogi came calling at his laboratory with bold claims about being able to lower their blood pressure using transcendental meditation (TM), the technique developed and recently imported to the West from India by their guru. "Why are you fooling around with monkeys?" they demanded to know. "Study us!" At first Benson turned them away with a polite "no thank you." His superiors had already warned him that his career was in jeopardy because of his studies of stress and white coat hypertension; to start investigating meditation would surely put him even further beyond the pale. But the TM enthusiasts were very persistent. "They wouldn't go away—they insisted on being studied," Benson told me. Eventually he relented. What could be the harm of a little preliminary investigation?

First he arranged a meeting with the Maharishi, who graciously agreed that his organization would cooperate in the research even if the results looked like they were going to be detrimental to his movement, and then he applied for ethical approval from the Harvard Committee on Human Studies.[3] Having received the blessings of the spiritual and temporal authorities, Benson and his colleagues set up a series of experiments. The TM enthusiasts sat in a chair and were wired up to various instruments for measuring their blood pressure, breathing rate, rectal temperature, and the amount of oxygen and other chemicals in their blood. They also wore an "electrode cap"—an array of sensors applied to the scalp—which monitored the electrical activity of their brain. Each session lasted around ninety minutes. The volunteers were given thirty minutes to get used to the presence of the equipment—some of it was a little invasive—then, as the measurements began, they were instructed to close their eyes and sit quietly, allowing their minds to wander, for twenty minutes. They were then told to meditate for the next twenty minutes, and finally to allow their minds to wander for the remaining twenty minutes. To avoid skewing the data, they were asked to keep their eyes closed for the whole sixty minutes of data collection and to avoid changing their posture.

"There were dramatic physiologic changes," recalled Benson, and even after the passage of half a century, I could hear the excitement of this discovery in his voice. As soon as the subjects started to meditate, their breathing rate and oxygen consumption plummeted and their heart rate slowed, indicating an abrupt decrease in metabolic rate. Their muscles relaxed, measured as a decrease in the amount of lactate circulating in their blood. At the same time, slow electrical oscillations in their brains known as alpha waves increased in intensity. The pattern of changes was nothing like that seen in sleeping humans or hibernating animals—two other states where the rate of metabolism falls. This was something entirely different. Ironically, the one thing that didn't change was the healthy young volunteers' blood pressure, which had been the original focus of the research. It was already low before the experiments and stayed low both during them and after.

What seemed to be happening was a reversal of the physiological effects that accompany the fight-or-flight response—the body's automatic reaction to dangerous situations such as the sight of a predator or rival, which prepares us to either attack or flee. The fight-or-flight response is orchestrated by the "sympathetic nervous system," part of the body's autonomic control system, and triggers changes including faster breathing and heart rates, increased blood pressure, blood glucose levels, and muscle tension. To experience stress, whether it is caused by an exam, a job interview, or a heated argument, is to experience the fight-or-flight response in action. An evolutionarily ancient system, it primes an animal's body for the vigorous muscular activity needed to battle for its life. The response is triggered in the brain by twin almond-shaped structures called the amygdalae (one in each hemisphere), which are intimately involved in fear responses, and leads to the release of the hormones epinephrine and norepinephrine (also called adrenaline and noradrenaline) into the bloodstream. These in turn cause dramatic physiological changes throughout the body.

Benson knew that when the danger has passed, a complementary network known as the "parasympathetic nervous system" swings into action to restore the body to a state more suitable for less vigorous but nonetheless vital activities such as feeding and grooming. He reasoned that people practicing transcendental meditation must be able to evoke these changes at will by turning up the activity of their parasympathetic nervous system, creating a feeling of profound calm and counteracting the effects of stress. Benson labeled it "the relaxation response," because the physiological effects were the polar opposite of those caused by the stress or fight-or-flight response. Through an extraordinary coincidence, half a century earlier, in the very same laboratory at Harvard where

Benson and his fellow researchers identified the relaxation response, the physiologist Walter Bradford Cannon had identified the fight-or-flight response.[4]

The Beatles, who had been experimenting with the psychedelic drug LSD, were reputedly drawn to TM because they thought it might provide a nonchemical ticket to altered states of consciousness.[5] The Maharishi developed the technique in India in the 1950s, drawing upon ancient Hindu practices, and brought it to America and the UK on a world tour in 1959. Responding to questions in television interviews, he would often dissolve into laughter, earning him the not entirely complimentary nickname "the giggling guru." The Beatles first met him in London in 1967—a year before Benson began his studies of the relaxation response—and the band even traveled to his ashram in Rishikesh, India, to sit at his feet, though they would later become disenchanted with their erstwhile guru and leave sooner than planned. Paul McCartney has said that the lyrics of "The Fool on the Hill" ("Day after day, alone on a hill, the man with the foolish grin is keeping perfectly still. . . .")—written shortly after they first met him—were inspired by the Maharishi.[6]

However foolish he may have appeared to onlookers, it turns out the giggling guru was evoking the relaxation response. In the terminology of its practitioners, the purpose of TM is to "transcend thinking" in order to achieve a state of restfully alert consciousness. People who want to learn the technique pay a fee to a certified teacher, who interviews them and explains the philosophy before issuing a mantra that is theirs alone and which they are not meant to utter aloud, let alone divulge to anyone. To meditate, they sit comfortably in some quiet place without adopting any special yoga posture, close their eyes, and repeat the mantra silently over and over again. If they become aware that they have been distracted by thoughts, they gently return their attention to the mantra. They do this for twenty minutes twice a day.

In the years following his identification of the relaxation response, Benson conducted further research to discover what exactly it was about TM that triggered these physiological changes. "If the relaxation response was the opposite of the fight-or-flight response, there should be other ways of evoking it," he reasoned. "There's more than one way to evoke stress, and this is the opposite." He started to investigate other forms of meditation. Far from being unique to TM, he found that focusing attention exclusively on a body movement (as in yoga), the breath, a word, a sound, or a repetitive prayer triggered exactly the same physiological response in research subjects. There were two common factors in all these practices. The first seemed to be that they stemmed the cascade of regular thought; the second was that when thoughts wandered into the mind, the meditator accepted them impassively but quickly returned his or her attention to whatever it was they were repeating. "What these two steps do is break the chain of everyday thinking," Benson told me. "That's fundamental to all these practices." He and his colleagues then spent several years reviewing religious and secular literature from a wide variety of traditions to see whether the same principle had been described before. They didn't have to look very hard. "It was striking, because every single culture of humankind had been doing this—different repetitions, different prayers, different movements—for millennia. It started in India with Yoga, and then we found it in Judaism and Christianity going back to the time of the Desert Fathers. We found it in Zen and Shinto Buddhism. In other words, it was universal."

This conclusion didn't go down well with the TM movement,

and when Benson went on to develop a standardized routine for evoking the relaxation response that involved repeating not a personalized, mystical mantra but the plain old English word *one*— well, that was the final straw. The Maharishi and his movement had always been grateful to Benson for establishing the scientific validity of transcendental meditation. "I had become a bit of a hero for them," he told me, somewhat wistfully. "But then when I discovered there were other ways to evoke the same physiologic changes they became quite upset." Benson said he owed a huge debt of gratitude to the stubborn young TM enthusiasts who had come knocking at his door in 1968 for setting his career off on such a fascinating trajectory, but, like John, Paul, George, and Ringo before him, the cardiologist was destined to fall out with the Maharishi. The TM movement, which was now becoming extremely wealthy in the US, began to fund its own research, eventually making this one of the most studied forms of meditation. Meanwhile Benson and his team were finding preliminary evidence that the relaxation response can counteract the adverse clinical effects of stress in a wide range of disorders—he reeled off a long list to me—hypertension, cardiac arrhythmia, heart attack, strokes, diabetes, all kinds of pain, headaches, premenstrual tension, inflammatory bowel disease . . . There was even evidence it could coax young college students into reducing their recreational drug use.

Nonetheless, decades would pass before the medical establishment accepted the health benefits of a quasi-spiritual practice such as TM. For many years, like Newton's clandestine dabblings in alchemy, Benson was forced to pursue his meditation research quietly, in parallel to more conventional work. "I had to conduct two careers at that time: one as a cardiologist and the other as 'my crazy thing,'" he said. At one point he came close to being thrown out of Harvard Medical School because his boss thought it was unforgivable for a doctor to be wasting time on so-called hippie nonsense. Benson appealed to the dean, Robert Ebert, who put a stop to any talk of dismissal with the memorable phrase: "If Harvard can't take occasional chances, who can!" Some fifty years later, the relaxation response (elicited by mindfulness meditation and TM, among many other techniques) has gone mainstream. A study published in the *American Journal of Hypertension*, for example, suggests that by reducing levels of psychological distress, TM can lower blood pressure in young people at risk of developing hypertension.[7] And in 2013, the American Heart Association tentatively approved the use of a range of meditation techniques as complements to conventional antihypertensive treatments.[8]

"Interrupting the train of regular thoughts," as Benson puts it, seems to be the universal key that unlocks the relaxation response, though it has remained something of a mystery how its beneficial clinical effects are mediated at the molecular level. Genetic research is now starting to provide some tantalizing clues. Already well into his eighth decade, Benson contributed to two studies published in 2008 and 2013 that revealed a wide range of changes in gene expression (the activity of particular genes) that are associated with the response.[9, 10] In the first study his team compared gene expression across the entire genome in two groups of people: those who had no prior meditation experience and those who had spent between four and twenty years practicing a technique known to elicit the response. The researchers gave the first group eight weeks of relaxation-response training, taking blood samples before and after to compare gene expression. Meditation in both groups appeared to suppress the long-term activity of genes involved in inflammation and stress responses, and crank up that of genes responsible for efficient energy metabolism, insulin secretion, and DNA repair. Some of the changes were more pronounced in people who had been practicing for years.

In the second study, they found that many changes in gene expression were triggered extremely rapidly by meditation, showing up on tests immediately after a session. These activity levels were sustained or had even increased fifteen minutes later and were greatest among experienced meditators. Again, there was a boost in the expression of genes involved in energy metabolism, DNA repair, and insulin secretion, and decreased expression of genes involved in inflammation and stress responses.

Benson believes the relaxation response reduces oxidative stress at the cellular level and tames the inflammation associated with so many illnesses known to be exacerbated by chronic stress, including hypertension, anxiety, insomnia, diabetes, and rheumatoid arthritis. The overall pattern of genetic changes in experienced meditators seems to be the same regardless of the technique used—whether it's qigong, mindfulness, TM, yoga, or repetitive prayer. The research also suggested that, in the long term, the relaxation response may combat cell aging by promoting the repair and maintenance of the stretches of expendable DNA at the ends of chromosomes known as telomeres, which act like the stiff plastic sheaths that prevent the ends of shoelaces from fraying. In order to divide, a cell must duplicate all its chromosomes, but every time this happens, some of the DNA in its telomeres is lost. Eventually, in cells that have to divide many times in the course of a lifetime, the telomeres wear down to such an extent that the integrity of the genes carried by the chromosomes is threatened. To protect the body as a whole, these cells stop dividing and become senescent, and may eventually undergo apoptosis—cell suicide. So, in most cells, the telomeres seem to act like lit fuses, steadily burning down until—bang!—it's all over.

Telomere length thus provides a measure of a cell's biological age and, on average, people with shorter telomeres seem to have lower life expectancy.[11] There is also solid evidence that working in a stressful environment shortens telomeres and accelerates aging.[12, 13] By promoting the activity of a gene that makes telomerase—the enzyme that rebuilds telomeres—the relaxation response may combat the effects of aging at the cellular level. Some very preliminary research even suggests that going on an intensive meditation retreat could slow this stress-related shortening of your telomeres by boosting telomerase activity.

The snowcapped peaks and dense pine forests of the Colorado Rockies provide a stunning backdrop for meditation retreats at the remote Shambhala Mountain Center. For three months in 2007, this was the setting for a unique scientific experiment conducted by Tonya Jacobs of the Center for Mind and Brain at the University of California, Davis, and her colleagues.[14] Sixty men and women were randomly assigned to two groups. Thirty of them stayed at the center, where they meditated for around six hours every day and received expert training in a variety of contemplative techniques, and thirty were put on a waiting list for the retreat. This second group of people, which was similar in terms of its male/female ratio and age profile, average body mass index, and prior meditation experience, carried on with their regular lives in the outside world and acted as controls for the experiment. At the end of the three months, blood samples were taken from both the experimental subjects and the controls at an on-site laboratory in the basement of the retreat center, so that the activity of the telomerase in their immune cells could be measured. Telomerase activity was found to be significantly higher in the people who had been on the retreat compared with the controls.

The authors concede their study had some limitations. For example, the number of participants was relatively small, and it's hard to know whether the increased telomerase activity was a result of all the meditation they were doing or simply because they had been removed from the stresses and strains of their everyday lives. (A similar difference in enzyme activity might have been seen if half of the subjects had been sent to a sunny beach somewhere for a relaxing holiday.) Participants filled out questionnaires before and after the retreat to gauge their levels of mindfulness and well-being, which did seem to indicate the increases in telomerase activity were mediated by improvements in these measures, but psychological questionnaires of this kind are notoriously unreliable. Nevertheless, combined with the evidence from genetic research, the study provides a tantalizing hint that the relaxation response—elicited by meditation—might slow cellular aging by ramping up the activity of telomerase. More research using "active control groups"—which pit the intervention being investigated against an equivalent activity—will be needed before doctors can recommend meditation as a way to keep the Grim Reaper at bay.

What is certain is that the relaxation response evokes a blissful sense of tranquillity. Benson argues that meditation achieves its stress-busting effects by interrupting the train of normal, everyday thoughts. He believes that in the distant evolutionary past of our species, when our ancestors were living on the African savannah, we paid a high price for our increasingly sophisticated brains because we were now capable of *thinking* about the dangers that threatened our existence. So, whereas creatures with less highly evolved cognitive powers were able to get on with foraging and feeding after the threat from a predator or rival had passed, humans didn't find it so easy. Their overactive imaginations kept their bodies primed for action, with all the damaging health effects this entailed. "It's not just having a saber-toothed tiger in front of you, it's the *thought* of the saber-toothed tiger," Benson told me. "It's the thought of someone or something injuring you that evokes the fight-or-flight response."

This would seem to suggest that the downside of having evolved a bigger brain is chronic stress. But there was an antidote available to anyone fortunate enough to stumble across it—a mental trick that could be used to rapidly restore one's physiology to a more relaxed, baseline state. "The many side effects of stress include anxiety, depression and anger, but people found by trial and error that if they adopted a certain posture or breathed in a certain way they broke the train of everyday thought that was evoking the fight-or-flight response . . . and they felt good!" said Benson. He believes the relaxation response is the foundation of all forms of meditation and the common thread that runs through many of the spiritual experiences described in the literature of the world's religions. "The ancients were right. All we've done is put it in the language of the day—science." His own outlook remains that of a scientist unaffiliated with any religion. The relaxation response, he told me firmly, is simply a way to temper the stress response. In fact, in the early years of his career, he was so anxious to maintain his objectivity he didn't attempt to elicit the response himself. "I was fearful of being accused of being a 'true believer.' But now I'm older and I need it." He sits down to meditate for twenty minutes twice a day before mealtimes.

It seems the relaxation response is so natural that even the eight-year-old Prince Siddhārtha managed to discover it for himself. The first step toward enlightenment is a calm mind, but it is only the beginning. There must have been more to his childhood experience, because as an adult he would describe it succinctly as "rapture and pleasure born from seclusion, *accompanied by directed thought and evaluation.*"[15] This translation hints that there was also *mindfulness,* which is the focus for much of the rest of this book. An Indian ascetic living in the fifth century BCE wouldn't have raised an eyebrow at the mindfulness element of the experience, but "rapture and pleasure" were surely anathema. Nevertheless, the homeless Siddhārtha reasoned that there was nothing wrong with enjoyment—even ecstasy—provided it had been uncoupled from the cravings of the body and the mind was firmly under control. "Why am I afraid of that pleasure that has nothing to do with sensuality, nothing to do with unskillful mental qualities?" he thought. Was there a middle way between the extremes of self-denial and sensory indulgence?

He knew he couldn't attain this happy, balanced state while his body was racked by hunger. One story has him taking his begging bowl to a nearby village where a brahmin girl offered him a meal of rice milk and another villager gave him some kusha grass to use as a mat during his final effort to attain enlightenment. When they found out, his five followers were disgusted at these indulgences. "Gautama the contemplative is living luxuriously!" they exclaimed. "He has abandoned his exertion and is backsliding into abundance!"[16] They deserted him, believing he had already failed in his quest.

But, as evening turned to night, Siddhārtha settled down to meditate cross-legged with his back to the fig tree and the setting sun. He vowed not to get up until he had realized his goal.[17] "Let my skin and sinews and bones dry up, together with all the flesh and blood of my body . . . I will not stir from this seat until I have attained the supreme and absolute wisdom."

Guided Meditation: Only the Breath

It's in the nature of the mind to wander, which may be why we hardly ever notice it happening. By training yourself to focus on a simple, predictable, and repetitive stimulus, such as the breath or a mantra—and retrieving your attention whenever it strays from the chosen target—you become better at noticing. Think of meditation as a gentle workout for your brain's attention circuits: with practice it will become easier and you will be able to focus for longer and longer periods of time.

Like any "focused attention" or "concentration" meditation, attending to the breath evokes the body's relaxation response by calming the mind and aligning it with what's happening in the present moment. Your only responsibility will be to concentrate on the sensation of each inhalation and exhalation, noting when your attention has wandered (which it will) and patiently returning it to the breath. That's it. Rather than striving to achieve a goal, which is what we spend most of our waking hours doing, consider this a welcome opportunity to stop and simply *be*.

Make meditation as much a part of your daily routine as brushing your teeth. Set aside five minutes, ten minutes, or half an hour, depending on your circumstances and experience; it doesn't really matter. The important thing is to practice regularly. Choose a quiet time when you are unlikely to be disturbed. Many people find first thing in the morning before breakfast works best, but if you find yourself struggling with drowsiness or hunger, try practicing after a light break-fast (sugar-free to avoid ramping up your insulin levels, which will make you lethargic).

Eventually you should aim to practice sitting cross-legged on a cushion on the floor, which is the ideal posture for stay-ing awake and alert, but if you are new to meditation or have a physical disability, a straight-backed chair is fine. Wear loose-fitting clothing that won't pinch or restrict the circu-lation. Five minutes' meditation is enough to start, but you can extend this as you become more skilled. Set the timer on your phone, selecting a gentle rather than a shrill alarm and turning off the vibration. Use the phone's "silent" or "air-craft" mode to avoid any electronic interruptions.

Your posture should be erect and balanced, with your head, neck, and back aligned but not stiff. The aim is to be relaxed, alert, and dignified, reflecting your desired state of mind. Experiment until you find a position you can sustain comfortably. If you're in a chair, sit with your legs uncrossed and the soles of your feet flat on the floor.

Take a few slow, deep breaths. Relax your shoulders and facial muscles. Notice the points of contact between your body and the floor, cushion, or chair. Now focus on the sensa-tion of your breath at your nostrils as you breathe in and out. Don't try to control or change the breath in any way, merely observe it. When you realize that your mind has wandered from the breath and you have started planning, analyzing, or musing about something, note what has happened and gently escort your attention back to the breath. If the thought seems important or pressing, you can tell yourself, "I'll deal with that later." Remember, everyone's mind wanders. Simply note what has happened without blaming yourself.

Restrict your attention to each inhalation and exhalation as it comes, nothing more. If you find yourself speculating how long it will be before the alarm goes off, answer this thought with "I'm only interested in this present moment of experience." Focusing on the breath is your only responsibility, so let go of everything else.

If you are having trouble staying focused, use the mantra *Buddho* ("knowing" or "conscious awareness"). Say "Bud-" during the inhalation, "-dho" during the exhalation. Make the mantra fully conscious rather than an automatic repetition. Try visualizing the spelling as you say it.

Alternatively, imagine a peaceful setting as you breathe in and out. It could be anywhere. You could be sitting near the mouth of a cave halfway up a mountain, looking out at the blue sky through the opening as if the cave were the cavity of your own nose. You could be in a boat, sailing a straight course on a wide, calm lake, the gentle pressure exerted with a hand on the tiller representing the steady, sustained effort of focusing on your breath.

When the alarm goes off, open your eyes, stretch, and give yourself a few moments. There's no hurry to get up. And remember, mindfulness is a state of attentiveness to the present moment you can bring to everything and anything you do. It's not just about meditation.

CHAPTER THREE

THE CLOUD OF UNKNOWING

All beings by nature are Buddha, as ice by nature is water. Apart from water there is no ice; apart from beings, no Buddha.

—Hakuin Zenji, "Song of Zazen," translated by Robert Aitken

The five ascetics are gathered around the shattered stump of a felled tree. As the sun climbs higher in the sky above the park, the heat becomes almost unbearable. Even the air burns in the nose and throat, bitter with smoke drifting from dozens of sacrificial fires. One of the five sits cross-legged in the red dust trying to meditate, wedged upright among the rudder-like roots of the felled tree, while the others are sitting on freshly cut logs that the woodcutter left scattered about the jagged stump. They stare listlessly at the ground or up at the burning sky. Their garments are a motley patchwork stitched together from scraps of cloth, tree bark, and the wings of dead birds. They seem oblivious to the flies—or perhaps they have simply chosen to endure them.

An atmosphere of sadness and defeat hangs about the heads of the five emaciated men living rough in the deer park at Isipatana, about six miles northeast of Varanasi. One of them is much older than the others. This is Kondañña, one of eight brahmins summoned to King Suddhodana's court thirty-five years earlier and the only one who predicted unequivocally that the infant Siddhārtha would become a buddha.[1] He must now be feeling the sting of his former teacher's failure even more keenly than these others— Bhaddiya, Vappa, Mahānāma, Assaji—all sons of brahmins who went to the palace all those years ago to bestow their blessings and pay their respects. Like Kondañña, they have given up families, comfortable homes, and safe careers in the hereditary priesthood to walk this austere path, which has led nowhere.

Nobody "invented" meditation. The chances are that for tens of thousands of years, humans everywhere have been sinking unawares into this altered state of consciousness. All it takes to evoke the relaxation response—the gateway to all forms of meditation— is to focus exclusively on a repetitive stimulus or movement long enough to break the stream of ordinary thought. In our deep prehistoric past, hunter-gatherers staring into the flames of their fires may have fallen into a meditative state, found the experience profoundly calming, and learned to do it at will. "That was probably one of the earliest human meditative experiences," says Jon Kabat-Zinn. "Anyone who has ever sat around a fire at night in the wilderness will know that after a while the talk dies down and everyone winds up gazing into the flames—very quiet, very still, very awake, very focused."

Kabat-Zinn began attending meditation retreats in 1965 at the age of twenty-one and—while continuing to pursue his studies as a molecular biologist at MIT—trained under teachers including Philip Kapleau, author of *The Three Pillars of Zen*, and Korean Zen master Seung Sahn. "Meditation nurtured something in me that I'd been looking for my whole life," he tells me. "It was not only a complement to my scientific and intellectual development, it was also deeply satisfying on a personal level." He doesn't view it as an inherently religious or mystical practice, however, simply as a way to cultivate wisdom through greater awareness and acceptance of what is happening from moment to moment. "We're not talking about some special magical state," he says. "We're talking about pure awareness."

Of course, the idea that hunter-gatherers meditated is speculative, just as we'll never know for sure when humans developed language or a sense of humor. Unlike walking upright or hunting, for example, meditation hasn't left any traces in the prehistoric record—there are no telltale fossils or tools. The subjects of the evocative cave art created by our ancestors were almost always animals, and in the rare instances where people have been depicted, they are usually hunting them. So there's no way to know when meditation became formalized as part of religious or shamanistic ritual. The oldest physical evidence that people had learned to meditate doesn't appear for tens of thousands of years after the debut of anatomically modern humans in the fossil record, some 200,000 years ago. It comes in the form of a carved stone seal dated to around 2500 BCE found among the archaeological remains of Mohenjo Daro, one of the largest cities of the Indus

Valley Civilization, in what is now the Sindh Province of Pakistan. The seal—which may well have been used as a stamp of priestly authority—depicts a seated, godlike figure wearing a tall headdress made from a pair of buffalo horns. He is sitting on a dais in a recognizable yoga position, legs bent double with the heels of his feet together, his arms outstretched with the hands resting on his knees. At his sides are four animals—an elephant, a rhinoceros, a buffalo, and a tiger—and beneath the dais on which he sits are two antelope or ibex. Some archaeologists have claimed the seated figure is the Vedic god Rudra, a forerunner of the Hindu deity Shiva, sometimes depicted as an omniscient Yogi, though this interpretation has been contested.[2] God or human, he appears to be meditating.

The earliest written references to meditative practices are found in the most ancient of all Indian scriptures, known as the Vedas, which date back to around 1500 BCE, with more detailed guidance provided much later in the Upanishads, the Bhagavad Gita, and the Yoga sutras. These describe the classic yogic discipline known as *pranava*, which involves the constant repetition with each breath of "Om"—the first or ultimate sound that is said to have brought the whole world into existence.[3] Repeating this mantra is supposed to help the meditator transcend suffering by uniting Atman, his or her "true Self" or soul, with Brahman, the "absolute reality" or universal consciousness. In India's pre-Hindu, Vedic religion, Brahman was the spiritual force the priests were said to channel through their animal sacrifices and chants, which may explain why they would later come to be known as brahmins.

In the classic of Hindu literature, the Bhagavad Gita, a fictional dialogue between the deity Krishna and his disciple Prince

Arjuna written as early as the fifth or fourth century BCE, Krishna provides what may be the first meditation self-help guide:[4]

> *Those who aspire to the state of yoga should seek the Self in inner solitude through meditation. With body and mind controlled they should constantly practise one-pointedness, free from expectations and from attachment to material possessions.*
>
> *Select a clean spot, neither too high nor too low, and seat yourself firmly on a cloth, a deerskin and kusha grass. Then, once seated, strive to still your thoughts. Make your mind one-pointed in meditation, and your heart will be purified.*
>
> *Hold your body, head, and neck firmly in a straight line, and keep your eyes from wandering. With all fears dissolved in the peace of the Self and all actions dedicated to Brahman, controlling the mind and fixing it on me, sit in meditation with me as your only goal. With senses and mind constantly controlled through meditation, united with the Self within, an aspirant attains nirvana, the state of abiding joy and peace in me.*

In the Yoga sutras, compiled by Patañjali in the fourth century CE, some of the more concrete benefits of repeating the word *Om* are outlined: "Mental pain, despair, nervousness and agitation are the symptoms of a distracted condition of mind. For removing these obstacles, there should be the constant practice of the one principle, the repetition and cultivation of Om."[5] Transcendental meditation, with its personalized mantras, is the direct descendant of this practice and clearly works on the same principle. You could also argue that when Herbert Benson formulated his simple in-structions for eliciting the relaxation response, he was simply re-placing *Om* with *One*. His research had revealed that sitting in a quiet place with your eyes closed, silently repeating a word of one syllable for about twenty minutes, broke the train of regular thought and in the process lowered blood pressure, slowed heart rate and breathing, and elicited a profound sense of calm. Benson argues that regardless of whether it is *Om, One,* or a more elaborate mantra, stripped of the religious language what remains is a way to evoke the physiological flip side of the fight-or-flight response: an antidote to the stresses of everyday life.

Ajahn Chah, the Thai Forest monk who brought this Buddhist tradition to the West in the 1970s, recommended using the word *Buddho* ("knowing" or "conscious awareness") for this purpose.[6] He explained it in a talk to newly ordained monks in 1978:

> *Just keep breathing in and out like this. Don't be inter-ested in anything else. It doesn't matter even if someone is standing on their head with their ass in the air. Don't pay it any attention. Just stay with the in-breath and the out-breath . . . "Bud-" on the in-breath; "-dho" on the out-breath. Just stay with the breath in this way until you are aware of the in-breath and aware of the out-breath. . . . Be aware in this way until the mind is peaceful, without irritation, without agitation, merely the breath going out and coming in.*

Meditative practices have played an important but less well-known role in Christianity, Judaism, and Islam. In his guidebook for Christian contemplatives, *The Cloud of Unknowing*, an anonymous fourteenth-century mystic claimed that during meditation, "the soul is made one with God." This union was seen as the highest achievement of which human beings are capable.[7] The author, probably an English Carthusian monk, wrote that before the Fall, humans were closer to God as a result of meditation. "This is the work in which humanity would have continued if we had never sinned, and it is the work for which we were made and everything was made for us to help and further us in it, and by it we shall once more be restored." He conceded that a Christian could worship God by performing charitable works, but he was clearly convinced that a life of contemplation would bring him closer to his maker. "This is the work of the soul that pleases God most. All the saints and angels rejoice in this work, and hasten to help it with all their might. All the devils are driven crazy when you do this, and try to frustrate it in all ways they can."

The monk wrote that the work of contemplation was like entering a "cloud of unknowing" between the meditator and God that could not be penetrated by reason. You had to do your best to stay in this dark cloud for as long as you could, "for if you are ever to feel or see him, so far as it is possible in this life, it must always be in this cloud and this darkness." Christian monks, nuns, and hermits would cut themselves off from the outside world to devote themselves to attaining this mystical union of the soul with God, and they did so in a manner that would be perfectly familiar to a Yogi trying to realize the unity of his true Self, Atman, and the universal consciousness of Brahman. It was hard work for a Christian contemplative (and not helped by all those devils), but there were "tricks and wiles and secret stratagems of spiritual technique" that could help suppress your mundane everyday thoughts. In accord with Benson's guidelines for eliciting the relaxation response, the author of *The Cloud of Unknowing* recommended an attitude of impassive acceptance toward distracting thoughts. One should not engage with them emotionally. Another of the monk's stratagems, echoed in many descriptions of meditation in different religious traditions, involved a word that the contemplative must repeat over and over but without making any attempt to investigate its meaning:

> . . . *take only a short word of one syllable; that is better than one of two syllables, for the shorter it is, the better it agrees with the work of the spirit. A word of this kind is the word GOD or the word LOVE. Choose whichever you wish, or another as you please, whichever you prefer of one syllable, and fasten this word to your heart, so that it never parts from it, whatever happens.*
>
> *This word is to be your shield and spear, whether you ride in peace or in war. With this word you are to beat on the cloud and the darkness above you. With this word you are to hammer down every kind of thought beneath the cloud of forgetting; so if any thought forces itself on you to ask what you would have, answer it with no more than this one word. And if, in its greatest learning, it offers to expound that word and tell you its attributes, say that you wish to have it quite whole, and not analysed or explained.*

Francisco de Osuna, a sixteenth-century Spanish Franciscan friar, recommended in *The Third Spiritual Alphabet* that Christian contemplatives rebuff distracting thoughts with an emphatic "no." He too advised against intellectual analysis:[8]

I warn you against discussing the matter further in your mind; it would greatly disturb your recollection; to exam-ine into the matter would be a hindrance; therefore, shut the door with "no." You will know that the Lord will come and enter your soul if the doors, which are your senses, are closed . . . But you will answer that it would be wrong to say "no" to God and he alone is expected. But God comes in some other way of which you know nothing.

Anyone who has meditated, whether or not they believe in a supernatural authority, will relate to the description in *The Cloud of Unknowing* of brief moments of transcendence in which our consciousness springs "swiftly to God like a spark from a coal" or how once in a while we may even "forget the whole created world, suddenly and completely," before the gravitational pull of some thought or recollection brings us back to earth. "But what of that?" muses the anonymous writer. "For straight afterwards it rises again as suddenly as it did before." These spiritual experiences could be surprisingly pleasant, he says, and to a modern reader some of his descriptions sound almost trippy. As a foretaste of the heavenly reward, God would sometimes "set aflame the body of a devout servant of his, here in this life, with most wonderful sensations of sweetness and pleasure. Some of these do not come into the body from outside through the windows of our sense, but from inside, rising and springing from abundance of spiritual joy, and from true devotion in spirit."

He tells his readers that they shouldn't feel guilty about these heavenly sensations. "Such experience certainly cannot be evil if the deceptions of intellectual ingenuity and of uncontrolled strain-ing of the bodily heart are removed as I am teaching you, or in a better way if you know one." This has distinct echoes of Siddhārtha's musings on the pleasurable state of mind he experienced during his childhood meditation experience: "Why am I afraid of that pleasure that has nothing to do with sensuality, nothing to do with unskillful mental qualities?" This was the middle way between harsh self-denial and sensory indulgence that he wished to explore, much to the disgust of his fellow ascetics.

There are other clear parallels between Christian and Buddhist meditation. A Christian contemplative will often call to mind the suffering of Christ on the cross, just as Buddhist monks and nuns sometimes meditate on the suffering of humans and other beings to cultivate compassion, considered an essential component of enlightenment (the creatures writhing in the wake of King Suddhodana's plow are reputed to have evoked compassion in the heart of his son). And in a passage that to all intents and purposes is a call for greater mindfulness, the author of *The Cloud of Unknowing* encourages his readers to "take good care of time and how you spend it, for nothing is more precious than time. In one little time, however little it is, heaven may be won or lost." Cultivating awareness of the present moment—the cutting edge of experience—is also an aim of meditation in Jainism, which predated Buddhism in ancient India.

The breath is often used as a focus of attention in Jain meditation and many other religious traditions. In Judaism, to provide conditions favorable for prophecy and union with God, thirteenth-century kabbalist Abraham Abulafia recommended reciting the letters of holy names in Hebrew in a particular order, "to unseal the soul, to untie the knots which bind it," in conjunction with specific breathing rhythms and bodily postures. Gershom Scholem, a twentieth-century Israeli philosopher and historian, found the similarity with Indian contemplative practices so striking, he even went so far as to say Abulafia's teachings "represent but a Judaized version of that ancient spiritual technique which has found its classical expression in the practices of the Indian mystics who follow the system known as Yoga."[9]

One might think meditation has gone out of fashion in world religions other than Buddhism and Hinduism, but in recent years there has been something of a revival of a Christian contemplative practice dating back to the Desert Fathers, which often involves repetition of the Jesus Prayer ("Lord Jesus Christ, Son of God, have mercy on me, a sinner") or a single sacred word. And it's worth pointing out that the line between prayer and contemplation can be blurry. How does saying "Hail Mary, full of grace . . ." fifty times, for example, differ from the dogged repetition of a mantra? The words of the Hail Mary have a religious significance whereas a mantra need not mean anything, but after a few dozen repetitions, how deeply is one really thinking about the Blessed Virgin? What is certain is that repetition clears a space in the mind by warding off stray thoughts. The same will apply whether one is repeating *Om, One, No, Adoramus te Domine,* or *Hare Krishna.* In Roman Catholicism, fingering the beads of a rosary, one for each Hail Mary, helps sharpen the attention still further.[10] Prayer beads are also employed in Sufism, the wisdom tradition of Islam. Sufis use them during a contemplative practice known as *dhikr* to keep track of repetitions of certain phrases or the names of God for the purposes of spiritual purification and to attain union with the divine. One result—though not the immediate objective—is to foster greater psychological well-being by enhancing moment-to-moment awareness and transcending the everyday self. The eleventh-century Persian mystic Al-Ghazali says of dhikr: [11]

> *Let him reduce his heart to a state in which the existence of anything and its non-existence are the same to him. Then let him sit alone in some corner, limiting his religious duties to what is absolutely necessary, and not occupying himself either with reciting the Koran or considering its meaning or with books of religious traditions or with anything of the sort. And let him see to it that nothing save God most High enters his mind. Then, as he sits in solitude, let him not cease saying continuously with his tongue, "Allah, Allah," keeping his thought on it. . . .*
>
> *He has now laid himself bare to the breathings of that mercy, and nothing remains but to await what God will open to him, as God has done after this manner to prophets and saints. If he follows the above course, he may be sure that the light of the Real [ultimate reality or God] will shine out in his heart. At first unstable, like a flash of lightning, it turns and returns; though sometimes it hangs back. And if it returns, sometimes it abides and sometimes it is momentary. And if it abides, sometimes its abiding is long, and sometimes short.*

After the establishment of settled Muslim communities in India in the twelfth century CE, the subcontinent became a melting pot for the wisdom traditions of Sufism, Jainism, Hinduism, and Buddhism. Some of the meditative practices that emerged probably spread much further afield, perhaps even coloring some strands of kabbalah in Judaism and the Christian monastic tradition in early thirteenth-century Europe.[12] But one of the common threads that runs through all these spiritual traditions is their use of focused attention to transcend the ordinary self and other barriers to experiencing ultimate reality.

Another common element is the discipline required to overcome the natural urges of body and mind. Buddhists identify five "hindrances" that obstruct successful meditation: sensual desire, ill will, sleepiness, overexcitement or depression, and doubt. While a monastic with a personality dominated by ill will would be advised to meditate on loving-kindness, compassion, or sympathetic joy, one easily swayed by sensual desire would meditate on the ugliness of the body in its ten stages of putrefaction: bloated, livid, festering, cut-up, gnawed, scattered, hacked and scattered, bleeding, worm-infested, and skeletal. For contemplatives with a tendency to intellectualize—which would probably include people who read books about the science of enlightenment—mindfulness of breathing is strongly recommended.[13]

Over hundreds of years, some Buddhist traditions have developed highly involved, esoteric techniques for establishing the mental calm that is the prerequisite for insight meditation and realizing the true nature of reality. The Tibetan system known as Mahamudra specifies twenty-one meditative practices that must be carried out in sequence to focus the attention to an increasingly fine degree.[14] Early sessions involve simple exercises such as focusing on a pebble or a twig, visualizing the syllable *hum*, or following the breath. Later sessions start to train the meditator to deal with thoughts as they arise, first learning how to suspend them completely, but later learning how to simply observe them, "not cutting them off at all, yet not falling under their spell." Each practice introduces a new image to express the increasingly subtle refinements of attention that are required. For example, in Practice 15 "you will keep your mind as if you were spinning a thread, keeping an even tension on it. For if your contemplation is too tight, then it snaps; and if it is too loose, then you slip into indolence." In Practice 17, if visual images pop into your consciousness, you try to keep your mind "as if it were a child looking at the murals painted on a temple wall . . . you must neither enjoy them nor fear them and thus neither think they are important nor cling to them." By Practice 18, the way you deal with thoughts as they arise has become spontaneous, effortless, and your mind is like "an elephant being pricked by a pin . . . you feel your thoughts occur, but you yourself never cut them off nor react to them in any way." Once established in this state of perfect calm, protected by mental armor as sensitive yet impervious to harm as an elephant's hide, the meditator can begin the work of realizing the perfect enlightenment that will loose the bonds of suffering.

Among all the world's religious traditions, the unique contribution of Buddhism may have been to demonstrate that, if you wished, you could remove supernatural beings from the picture and start investigating the mind solely for the purpose of improving well-being. Having shown the gods the door, essentially what we were left with was psychology. According to this new perspective, our minds are fundamentally pure but have become contaminated by the "poisons" of craving, aversion, and delusion, which are said to be the roots of all suffering, from anxiety and fear to jealousy and depression. The first stage of meditation—calm—helps you resist these defilements and see them more clearly, while the second—insight—allows you to discern the true nature of the mind and start traveling the road leading to the final, perfect enlightenment.

For Siddhārtha Gautama, it wasn't enough to be engulfed by "the cloud of unknowing," as the Carthusian monk would describe it two millennia later, and wait patiently in the darkness for his soul to be united with God or the universal consciousness. To attain the final enlightenment, he must somehow pierce the cloud, once and for all.

Like an airborne mountain range, heavy black monsoon clouds roll in from the Indian Ocean far away to the southwest. They hide the sun and plunge the park into a premature twilight. Lightning in the east heralds a dull rumble of thunder, and at last, the first fat drops of the rainy season kick up the dust around the tree stump. Soon they're falling thick and fast. The long, dry summer is over, but none of the starving men shows any inclination to find shelter. They barely seem to notice the rain even as it soaks into their hair and beards. The meditating ascetic remains motionless in the dirt as the drops trickle in rivulets down his weather-worn face and tumble over the bark and feathers that pass for his clothes.

Kondañña is the first to see him, a distant figure walking toward them through the rain. For a long time the old man seems uninterested by this other being, until a flash of lightning freezes the raindrops in midair and illuminates him. Kondañña gasps and struggles to his feet, then rouses the others. After a moment's hesitation and disagreement, they hurry to meet Siddhārtha, greeting him as a friend. Something in his demeanor quells their former reservations. All has been forgiven, and just three words from their teacher are sufficient to convey the momentous thing that has happened to him since they parted two months ago: "I am awake."

CHAPTER FOUR

THE SECOND DART

So I too found the ancient path, the ancient trail travelled
by the Awakened Ones of old.
— *Saṃyutta Nikāya*, 12:65, translated
by Ñanamoli Thera

The downpour comes to an end almost as abruptly as it started, leaving the air brilliant and clean. They find a log for Siddhārtha to sit on and one of them somehow conjures up water and a towel so he can wash his feet after the long journey.[1] There *is* something different about him: they can see it in his eyes. He begins to explain his reasons for giving up the extreme form of asceticism they have all been practicing these past few years. It is perfectly true, he says, that to devote your life to sensual pleasure is "base, vulgar, common, ignoble, unprofitable" . . . but to devote yourself to the opposite extreme, to punishing the body until it becomes weak and useless, is just as ignoble and unprofitable. Once he had realized this, he resolved to develop a new approach that steered a middle course between the two extremes. This led to a sense of profound peace, to the insight and spiritual awakening they had all been seeking for so long.[2]

In keeping with this new middle way, they must eat. So, before he continues, Siddhārtha sends three of them into town for alms. While they are away he begins to teach the other two, and later, when the first group has come back with food, he instructs them while the others go for alms.[3] He teaches them the Four Noble Truths: suffering, its cause, its ending, and the path that leads toward its ending. He tells them they must recognize, investigate, and understand each of these truths for themselves if they want to realize enlightenment.[4] Suffering is a basic fact of existence, he says, and the cause is attachment to pleasures and the urge to be rid of anything unpleasant or painful. We are like deer caught fast in a trap, he says, waiting helplessly for the hunter to return. But we can free ourselves by following a path of wisdom, ethical conduct, and concentration—the Noble Eightfold Path. This will end suffering by liberating us from all cravings, aversion, and delusion.

The men and women living at Amaravati Buddhist Monastery on a hilltop in the Chiltern Hills of southeast England are direct spiritual descendants of those five ascetics who became the newly enlightened Buddha's first followers some twenty-five centuries ago. I had the pleasure of staying at the monastery in the summer of 2014, and it was a revelation.

Buddhist monks and nuns all over the world still subscribe to the Four Noble Truths and follow the Noble Eightfold Path (see Figure 1, page 77) that the Buddha is said to have outlined that day in the deer park at Isipatana (now called Sārnāth) in northeast India. They too chart a "middle way" between the extremes of sensory indulgence and punishing austerity, though to an outsider like myself, the monastic life looks anything but easy. Monks in the Theravada tradition of Sri Lanka, Thailand, Burma, Cambodia, and Laos must adhere to 227 rules (fully ordained nuns have somehow been landed with no fewer than 311 rules) that govern every aspect of their lives. Their only real possessions are three robes and an alms bowl. They eat just once or twice a day, always before midday, and they regularly fast, living for days on nothing but water (filtered to avoid killing any living creature, no matter how small). While the rest of us dream of a day when we'll want for nothing, their objective is to want nothing: they must remain celibate and penniless and may not take anything that hasn't first been offered to them. Chatter is discouraged, and many choose to observe "noble silence" much of the time. At Amaravati, which is the largest Theravadin monastery in Europe, the monastics don't listen to music, watch television, use mobile phones, or have access to the internet, and the only books in their library are about Buddhism and spirituality. Much of their waking lives is spent in meditation, mostly in isolation in their *kutis*—rustic huts in the woods that border the monastery.

Anyone visiting such a monastery for the first time will be struck by an apparent contradiction. How can the road to nibbāna/nirvana—popularly viewed as an almost magical realm where everything is just fine—be paved with so much physical and mental deprivation? In truth, the Buddha never promised anyone eternal happiness and freedom from pain; he simply pointed out a way to transcend suffering. A monastery is a boot camp for the mind, if you will, providing intensive training for those who aspire to the incomparable peace of enlightenment. Isolation and strict discipline help the monastics to loosen the grip of their worldly attachments, which, according to the Buddha, is essential to realize this goal.

THE FOUR NOBLE TRUTHS

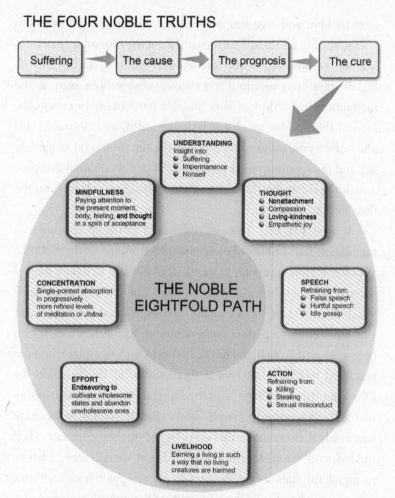

| Suffering | → | The cause | → | The prognosis | → | The cure |

UNDERSTANDING
Insight into:
- Suffering
- Impermanence
- Nonself

MINDFULNESS
Paying attention to
the present moment,
body, feeling, **and thought**
in a spirit of acceptance

THOUGHT
- Nonattachment
- Compassion
- Loving-kindness
- Empathetic joy

CONCENTRATION
Single-pointed absorption
in progressively
more refined levels
of meditation or *Jhāna*

THE NOBLE EIGHTFOLD PATH

SPEECH
Refraining from:
- False speech
- Hurtful speech
- Idle gossip

EFFORT
Endeavoring to
cultivate wholesome
states and abandon
unwholesome ones

ACTION
Refraining from:
- Killing
- Stealing
- Sexual misconduct

LIVELIHOOD
Earning a living in such
a way that no living
creatures are harmed

Figure 1. The Four Noble Truths and the Noble Eightfold Path.

As guests at Amaravati, we had to observe just eight basic precepts during our stay. We were obliged to show restraint in our behavior—no games, makeup, music, mobile phones, and so on—and to refrain from intentionally killing any living creature, from stealing, sexual activity, lying, "overindulgence in sleep," eating after midday, and ingesting intoxicating substances. We were also encouraged to play a full part in the life of the monastery by cleaning and helping out in the kitchen and grounds, and by attending the hour-long morning and evening observances, *puja*, in the meditation hall. Most of this was not particularly onerous—the food at the midday meal was tasty, plentiful, and nutritious, tea and coffee were allowed throughout the day, and a bit of gardening and dishwashing never hurt anyone—but meditation can pose a major problem for those not used to sitting cross-legged on the floor for long stretches.

At 5:00 A.M. each morning we would gather in the dimly lit meditation hall to kneel before the gilded image of the Buddha. When everyone was ready, a monk went forward to light a stick of incense on the simple, unadorned altar. Then, as the sweet scent filled the air, he led us in a chant: "To the Blessed One, the Lord, who fully attained perfect enlightenment; to the teaching which he expounded so well and to the Blessed One's disciples who have practiced well . . ." We bowed three times: to the Buddha, the Dhamma (his teachings), and the Sangha (the monastic community), often known as the Triple Gem of Buddhism. Once the chant had finished, everyone settled cross-legged on their cushions, a bell tinkled, and complete stillness fell. Almost straight away I began to regret my habit of meditating in an armchair back at home. I wasn't used to sitting bolt upright with my back unsupported, each of my long, middle-aged legs bent painfully at the knee like a pair of rusty old pliers. I don't think my body had assumed such a position for any length of time since I sat with the other fidgety kids on the floor during morning assembly at primary school. Soon my lower back, knees, and ankles were screeching in protest and my mind was in a turmoil over how long I would have to endure this pain before the bell rang to signal the end of the session. Thirty minutes later, when I finally gave in—uncrossing and stretching my legs and opening my eyes—I saw that the shaved domes of the monks' heads were as immobile and their backs as straight as they had been at the start, and yet they were somehow perfectly relaxed. They had entered a realm of stillness and composure to which I did not have access.

Jon Kabat-Zinn trained under several Zen Buddhist teachers in the 1960s and 1970s and remembers well the discomfort he felt during his early experiences of meditation. "When I started practicing on intensive retreats it was actually very painful," he tells me. "The first one I took we made a vow not to make any voluntary movements for periods of up to an hour and a half, two hours . . . I learned an awful lot about pain!" Retreats could last weeks, during which dedicated practitioners would sit cross-legged for more than eight hours each day. One of the most important lessons that Kabat-Zinn learned from this experience was that by changing one's mental attitude toward pain, its sting can be drawn. "There is a distinction between pain, which is part of the human condition, and suffering, which is something that we can compound when our minds become inflamed because we don't like what's happening." This is fundamental to the whole practice. The lesson stayed with him and in 1979, when he developed the Stress Reduction and Relaxation Program at the University of Massachusetts Medical Center in Worcester, he began applying it for the benefit of patients suffering from chronic pain that conventional treatments such as painkillers and surgery hadn't been able to shift. Many of these people had been experiencing debilitating pain for several years; their doctors had simply told them, "You're going to have to live with this." It was Kabat-Zinn's job to help them learn *how* to live with it.

Clearly there is more to our experience of pain than the alarm signals pinged to the brain by sensors scattered all over the body—in our skin, muscles, bones, joints, and internal organs. The way we *think* about the painful sensations and the emotions associated with such thoughts are just as important. The Buddha understood this intuitively and had a neat metaphor to explain the insight to his disciples.[5] "It is as if a man were pierced by a dart and, following the first piercing, he is hit by a second dart," he told them. "He worries and grieves, he laments, beats his breast, weeps and is distraught. So he experiences two kinds of feeling: a bodily and a mental feeling." Someone who hasn't been taught how to cope with painful sensations resists and resents them, the Buddha said. The only way he knows to escape the suffering is by distracting himself with sensory pleasures—which, as he taught in that first lesson in the deer park, come with dangers of their own attached. On the other hand, he said, "a well-taught, noble disciple, O monks, when he is touched by a painful feeling, he will not worry nor grieve and lament, he will not beat his breast and weep, nor will he be distraught." So, even if the first dart had found its mark, the second one could no longer hurt him.

This is not merely theoretical speculation. Examples abound of highly trained Buddhist monks enduring tremendous physical pain without flinching, the archetypal case being the self-immolation by fire of Quang Duc (Thích Quảng Đức) in Saigon on June 11, 1963, in protest against the discriminatory policies of South Vietnam's government. American photographer Malcolm Browne's extraordinary photograph, which was featured on the front pages of newspapers around the world, shows the sixty-seven-year-old monk immobile in the lotus position as his body was consumed by flames. While this is obviously an extreme example, it illustrates the mind's potential for enduring the most intense forms of pain imaginable. But what of the everyday pain experienced by more ordinary folk? Even in this era of high-tech analgesics and anesthesia, pain still causes a great deal of suffering. As many as a third of adults experience some kind of chronic pain, and this figure rises with increasing age to more than half of those over seventy-five.[6] Pain is associated with psychological distress and decreased quality of life, and it puts severe limits on everyday activities. In the United States alone, it has been estimated to cost a staggering $635 billion a year in medical treatment and lost productivity.[7]

In the hospital where Kabat-Zinn set up his clinic, doctors working in almost every department began to refer to him patients suffering from stress and anxiety related to a wide variety of conditions, not just chronic pain but also illnesses such as heart disease and cancer. He was essentially teaching mindfulness, without the religious and cultural overlays of Buddhism. The course lasted ten weeks, comprising homework and a two-hour session once a week during which patients were instructed in mindfulness meditation, the "body scan"—gradually sweeping their attention from their feet to their head—and hatha yoga postures. Mindfulness meditation involves first focusing one's attention on the sensation of breathing, as described in the second chapter (eliciting the calming "relaxation response"), then opening up the field of awareness to include physical sensations, thoughts, memories, and emotions as they arise. The meditator is taught to observe all these simply as passing mental events, without evaluating them or getting caught up in them. You can try this meditation by following the detailed instructions at the end of this chapter.

The secret Kabat-Zinn learned as a young man enduring grueling Zen meditation sessions is that if you can turn your full attention on pain without pursuing the emotional narratives that usually accompany it, you suffer much less as a consequence. It is a skill he still teaches to patients with chronic pain. "We don't just tell people, 'accept it and it will be okay,' " he says. "But paradoxically, when you begin to *befriend* your pain—move into it, embrace it, hold it in awareness—you begin to see that the suffering lies in thinking 'this is going to last forever' and 'it's destroyed my life, I'm never going to be better again.' That's not actually pain, those are just thoughts." He admits that this change in perspective is a challenging thing to ask of someone experiencing debilitating pain, but he says the key to reducing the suffering associated with an intense sensation is to turn toward it rather than try to run away from it. "That's where the rubber meets the road."

In 1982, Kabat-Zinn published results of the first clinical investigation into the efficacy of mindfulness meditation for easing chronic pain.[8] There were fifty-one patients in his study experiencing various types of pain, mostly lower back, neck and shoulder pain, and headache. Having completed the ten-week course, 65 percent saw their pain reduced by more than a third, when scored on a standard index that combines pain intensity and unpleasantness. In half of them, pain ratings had fallen by more than 50 percent. These changes in the patients' ability to cope with their pain were accompanied by significant improvements in mood and psychiatric symptoms.

Here was the first evidence that Kabat-Zinn's program, which is now known as Mindfulness-Based Stress Reduction, or MBSR, and generally lasts eight rather than ten weeks, had clinical benefits. In the ensuing three decades there have been thousands more studies with clear benefits established for pain, stress, and anxiety.[9, 10] The program has also been successfully tailored to prevent relapse in depression (see chapter 6, "Golden Slippers") and treat addiction (see chapter 7, "Fire Worshippers"). In all these diverse applications, the "two darts" approach is applied: patients are encouraged to recognize the distinction between the discomfort itself—whether that's the bodily sensations associated with stress and anxiety, or drug withdrawal—and the suffering we create around those sensations by our emotional reaction and the struggle to push them away. Mindfulness means accepting that some experiences are unpleasant without getting caught up in them emotionally or trying to push them away.

Kabat-Zinn has always believed that he is teaching people a life skill rather than providing a onetime treatment. "There is no age when it's not valuable to learn how to handle stress," he told me. Research has shown that chronic stress depresses our immune systems, worsens inflammation, affects nerve growth in the brain, and raises the risk of cardiovascular disease. "If you learn how to deal with stress and self-regulate—and the same is true of stress-related pain whether it's headaches, migraines or more serious things—then you increase your chances of staying healthy. And all that comes out of a greater awareness of your own mind and your own body and your own relationship to stress and pain, as well as what gives you pleasure in life and what gives you joy and satisfaction."

Since Kabat-Zinn developed MBSR in the 1980s, studies have investigated its potential worth in a wide variety of settings—from children taking exams to cancer patients going through the psychological trauma of their diagnosis and subsequent treatment. It is worth noting that in recent years, the methodological rigor of many of the early mindfulness studies has been called into question as the standards expected of psychological and clinical research in general have been raised and the bar for "proof" of clinical benefits is now set much higher. I explore some of these issues in chapter 11, "Mind Mirrors." At the same time, it should be recognized that a new approach to treatment takes time and resources to garner enough hard evidence of efficacy. The early studies were necessarily small and exploratory in nature.

From the start, Kabat-Zinn acknowledged the shortcomings of his pilot study in 1982 on the worth of MBSR for treating pain, including the lack of a control group against which to compare the treatment and potential biases in patients' reports of their pain (for example, they may have exaggerated the benefits of the program simply to be nice to their teacher). He also wrote that ideally in this kind of study an independent panel rather than the investigator himself should interpret some of the pain measures. Three years after the publication of this preliminary work, Kabat-Zinn did indeed publish a bigger study, carried out with the help of two independent collaborators at the hospital, in which he compared ninety chronic pain patients who took his mindfulness course with patients who continued to receive standard treatments.[11] He found statistically significant improvements in self-reported pain measures and reductions in anxiety and depression compared with standard treatments. By the end of the course, patients were also using less pain medication. Remarkably, the psychological improvements were still evident up to four years later, with the majority of patients reporting that they were keeping up their meditation practice. Interestingly, the *intensity* of their pain was unchanged from before the mindfulness program, but they were *coping* with it better. In other words, it was less likely to be dominating their lives.

Kabat-Zinn's research tends to involve mixed groups of patients with a variety of conditions. More recent studies are providing evidence that mindfulness can improve quality of life and reduce the discomfort associated with specific chronic pain conditions, including lower back pain, back/neck pain, arthritis, irritable bowel syndrome, and fibromyalgia.[12-15] Still, much clinical research remains to be done to establish just how good mindfulness is for easing the suffering associated with chronic pain and which types of patient might benefit most. The neuroscience of pain—with its complex interplay of alarm signals from sensors around the body, beliefs, emotions, and thoughts—is also still in its infancy, with many mysteries yet to be solved. Nonetheless, researchers are beginning to unpick how meditation affects the way pain perception is processed in the brain.

Suppose a time-traveling neuroscientist were to journey back to the fifth century BCE at regular intervals to scan Siddhārtha's brain, from those early years practicing meditation with his fellow ascetics through to his enlightenment. What would he or she discover and what might the findings tell us about pain and suffering? Lab-based studies suggest that even beginners—much like the twenty-nine-year-old Siddhārtha after he left his father's palace and began his spiritual quest—can dramatically change their experience of pain using mindfulness meditation. In 2011, researchers led by Fadel Zeidan at Wake Forest University School of Medicine in North Carolina found that men and women given just four sessions of mindfulness instruction, each lasting only twenty minutes, were able to reduce the unpleasantness of pain caused by a heat probe applied to their legs by 57 percent on average. And they reported that meditation reduced the intensity of their pain by 40 percent.[16] During the experiment, the subjects lay with their heads inside the doughnut-shaped magnet of an MRI scanner, which revealed that changes in the perceived intensity and unpleasantness of their pain were associated with distinct regions in their brains. Reduced *unpleasantness* correlated with increased activation of an area at the front of each hemisphere of the brain immediately above the eye sockets, in the orbitofrontal cortex, a part of the prefrontal cortex involved in assigning an emotional "charge" to incoming sensory data, which is vital for learning and decision making. Reduced pain *intensity* was associated with increased activity in an area further back on the inner surface of each hemisphere, known as the anterior cingulate cortex (ACC), which detects conflicts between tasks competing for attention and is also known to play a vital role in pain perception (see Figure 2, page 87). The cingulate cortex curls around the corpus callosum—the tract of nerve fibers connecting the two hemispheres of the brain—rather like a collar (*cingulum* is Latin for "belt" or "girdle" and *cortex* means the outer layer or rind of something).

Reduced pain intensity was also accompanied by greater activation in another cortical structure, called the insula, hidden deep inside the brain within each hemisphere, tucked into a fissure between three lobes. The insula is involved in the experience of emotion, awareness of internal bodily sensations and of their "salience"—in other words, how worthy of our attention they are amid all the other sensory data. Both the ACC and insula feature prominently in results from meditation brain research. It has even been suggested that between them they create the sense of conscious awareness [17]—an extraordinary claim whose significance I examine in more detail in chapter 10, "Wonderful and Marvelous."

Research suggests that the signature of pain in the brains of more experienced meditators is different from that of beginners.

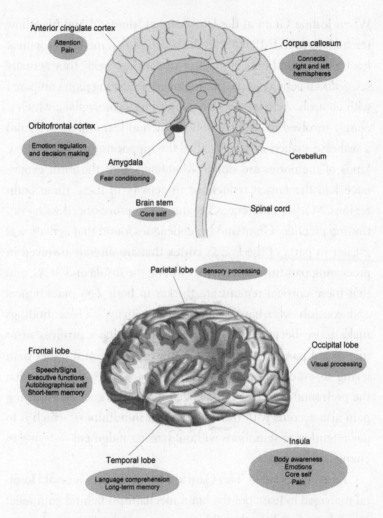

Figure 2. Drawing the sting. A short course of mindfulness training can help people cope better with chronic pain. Studies have found that increased activity in the orbitofrontal cortex of meditators is associated with decreased *unpleasantness* of pain; while increased insula and anterior cingulate cortex (ACC) activity is associated with decreased pain *intensity*. *Top*: the right cerebral cortex showing the midline between the two halves of the brain. *Bottom*: the outer surface of the left cerebral cortex with a section cut away to reveal the insula (a fold of cerebral cortex hidden deep in a fissure separating the temporal, frontal, and parietal lobes).

When Joshua Grant at the University of Montreal and his fellow researchers used fMRI to scan the brains of Zen meditators with at least a thousand hours of practice under their belts, they actually saw *reduced* activity in the prefrontal cortex during pain compared with controls. Activity was also reduced in the amygdala, which is closely involved in strong emotions such as fear and anxiety, and a seahorse-shaped structure called the hippocampus, where these kinds of memories are stored. Meditators with the most experience had the largest reductions in activity in these three brain regions.[18] Other differences became more pronounced with continuing practice. Grant and his colleagues found that activity was greater in parts of the brain's cortex that are directly involved in processing painful sensations, including the insula and ACC, and that these cortical regions are thicker in both Zen practitioners and controls who have lower pain sensitivity.[19] These findings make sense because, as we've seen, mindfulness involves turning toward pain, focusing one's whole attention on it rather than trying to evade or suppress it. Similarly, Grant's discovery that the prefrontal cortex of experienced meditators goes quiet during pain also accords with the stated aim of mindfulness, which is to pay attention to sensations without passing judgment or trying to change anything.

Researchers led by Tim Gard at Massachusetts General Hospital managed to tease out the brain mechanisms behind pain relief in meditation still further. They used an electrode applied to the forearm of volunteers to deliver focused electric shocks—which apparently feels like being poked with a pin—at random intervals. Compared with people who had never meditated, experienced meditators were able to reduce their anxiety levels by 29 percent as they were awaiting the painful stimulus. While they did this, activity in their ACC increased. When the pain arrived, they were able to use mindful acceptance to reduce its unpleasantness by 22 percent compared with the controls. They showed decreased activity in their prefrontal cortex and increased activation in the insula and secondary somatosensory cortex, where data from receptors all over the body for stimuli such as pressure, pain, and warmth are processed.[20] The researchers found that the better a meditator was at reducing the unpleasantness of the pain, the more strongly these areas were activated in his or her brain.

These changes in the brain neatly reflect the two distinct components of mindfulness meditation: "focused attention" and "open monitoring." Focusing attention on the breath first helps to create a sense of calm—combating anxieties about when the next stab of pain will come, and relaxing the muscles, which will lessen the pain when it comes. Its signature in the brain includes increased activation of the ACC (to recap, this region controls focusing and holding attention). By contrast, open monitoring involves tuning in to the sensory experience of pain (increased activation of the insula and somatosensory cortex), but with a completely nonjudgmental, accepting attitude (decreased prefrontal cortex activation). Neuroscientists call this type of regulation—also deployed during emotion regulation in experienced meditators (as we'll see in chapter 8)—"bottom-up processing." The rookie meditators given just four twenty-minute training sessions in Zeidan's study seem to have learned the easy trick of focusing their attention on the breath to reduce their anxiety levels, which would tend to lessen the emotional impact of pain. But in addition they seem to have attempted some cognitive or "top-down" control to lessen the pain.[21]

This strategy is not unique to meditation and involves reappraising the pain—"I can handle this," "It's not so bad," "A little pain can't do me any harm"—or distracting oneself by focusing on something else, whether that's another bodily sensation like the breath or a cerebrally demanding task such as mental arithmetic or planning the week's shopping.[22] There is also a placebo effect that comes from simply believing something is going to work—"This meditation business is powerful stuff."[23] Having faith in the efficacy of a pill, a medical procedure, a doctor, or a ritual is often enough to evoke real physical and psychological changes, and pain is particularly amenable to such effects. A placebo can ease pain either by triggering the release of endorphins, the body's own painkillers, or by sending a subliminal message to the brain to reappraise it as less unpleasant or harmful. However, a study by Zeidan and his colleagues in 2015 suggests that even in beginners, the pain relief provided by mindfulness meditation is more effective than that provided by a placebo alone, and they deploy different top-down cognitive control mechanisms.[24] This goes a long way toward rebuffing claims that meditation is just another way to evoke the placebo effect.

As we've seen, experienced meditators don't resort to any kind of top-down cognitive reappraisal strategy but instead open themselves up completely to the experience of pain without judging it or trying to change it. This attitude was at the heart of the Buddha's teaching about suffering. He taught that through mindfulness one could learn to view both pleasant and painful sensations with equanimity. The idea of "nonduality"—treating pleasure and pain, triumph and disaster alike—was important in early Hinduism, certainly by the time the Bhagavad Gita was written, perhaps as early as the fifth or fourth century BCE. In this epic tale of war and Yoga, Krishna teaches the concept to Prince Arjuna:[25]

> *When the senses contact sense objects, a person experiences cold or heat, pleasure or pain. These experiences are fleeting; they come and go. Bear them patiently, Arjuna. Those who are not affected by these changes, who are the same in pleasure and pain, are truly wise and fit for immortality.*

In the same vein, attachment to sensory experiences—selfish desire—came to be considered an important source of suffering by Hindus. To judge from this passage, the author of the Bhagavad Gita would have recognized and applauded Siddhārtha's first lesson about the perils of desire:[26]

> *When you keep thinking about sense objects, attachment comes. Attachment breeds desire, the lust of possession that burns to anger. Anger clouds the judgment; you can no longer learn from past mistakes. Lost is the power to choose between what is wise and what is unwise, and your life is utter waste. But when you move amidst the world of sense, free from attachment and aversion alike, there comes the peace in which all sorrows end, and you live in the wisdom of the Self.*

We can't know for certain whether these were originally Buddhist ideas, or whether they were already becoming widely acknowledged truths among spiritual seekers in the Ganges Plain in the sixth and fifth centuries BCE. This uncertainty leaves us to wonder what the Buddha taught his old friends in the deer park that so stirred them they were persuaded to abandon their path and accompany him on his. They had all been practicing the Yoga taught by sages such as Ālāra Kālāma and Uddaka Rāmaputta with a view to escaping saṃsāra, the painful cycle of birth, death, and rebirth. Short of building a time machine, we likely will never know what unique additions Siddhārtha made to the methods and beliefs about enlightenment that were already practiced among the renouncers. He would certainly place a greater emphasis on the power of loving-kindness and compassion to bring his followers closer to enlightenment. But some historians have argued that the key Buddhist innovation was the "open monitoring" component of mindfulness. According to this view, Buddhist meditation was about working *with* the mind, whereas the practice of meditation in Jainism and the Yoga of Vedic tradition was concerned with silencing the mind.[27] Like his fellow ascetics, Siddhārtha had been using brute force to suppress his senses and his mind's natural tendencies—by starving himself almost to death and even trying to stop his breathing—but this top-down strategy had failed miserably. His breakthrough came only with the realization that if he exposed the fetters of desire and pain to the intense spotlight of nonjudgmental, mindful awareness, they fell away. It seems the aging Kondañña already had an inkling of this, for when Siddhārtha had finished speaking, Kondañña replied with a powerful insight, which can be paraphrased as "Whatever arises in the mind's awareness will also cease." With these words he was not only acknowledging that all mental phenomena are impermanent, but also observing that they fade more quickly in the bright light of conscious awareness. The Buddha responded excitedly: "Kondañña knows! Kondañña truly knows!" His old friend was within touching distance of enlightenment.[28]

The realization of nonduality was certainly already established as a goal of meditation by the time the Bhagavad Gita was written, but Siddhārtha seems to have hit upon an effective way to achieve this happy state that didn't involve the devotee dragging his body to death's door. The brahmins probably wouldn't have had any problem with this new approach—they may even have welcomed it, though they balked at his establishment some years later of a Buddhist monastic order for women, his rejection of India's caste system, and his dispensing with the brahmin practice of sacrificing animals in holy fires. All this went against the established order. But Siddhārtha espoused another revolutionary idea that was much more unpalatable to them than any of these innovations. In the course of his inward investigations he had discovered something that—if you were to believe it, and most of us still don't—shattered their worldview and was regarded as heresy. He had discovered he didn't have a Self.

Guided Meditation: Open-Minded

A great way to return your mind to its "ground state," neither overexcited nor torpid, simply alert and open, is to become aware of the natural rhythm of the breath as you inhale and exhale (see "Guided Meditation: Only the Breath," page 56). This is focused attention, a prerequisite for the second stage of mindfulness meditation: insight.

Take up your position on a cushion or in a chair. Start by focusing on the sensation of the breath entering and leaving your body at the nostrils. Remember, you are observing your breathing rather than controlling it. Follow each inhalation and each exhalation from the start to the finish. Notice any slight gap between the in-breath and the out-breath.

Don't be hard on yourself if your mind wanders or you get distracted by a noise. This is all perfectly normal. Just remind yourself: "That's how the mind works," and return to the breath. With repetition, you will get better at noticing when you have lost focus and develop greater mindfulness of the present moment.

Now that you have quieted your mind, allow your attention to broaden. Whenever a positive or negative feeling arises, make it the focus of your meditation, noticing the bodily sensations associated with it: perhaps a tightness, the heart beating faster or slower, butterflies in the stomach, relaxed or tensed muscles. Whatever it is, address the feeling with friendly, objective curiosity. You could silently label whatever arises in the mind, for example: "There is anxiety," "There is calm," "There is joy," "There is boredom." Remember, everything is on the table, nothing is beneath your attention.

If you experience an ache or a pain, an itch or any other kind of discomfort, treat it in exactly the same way. Turn the spotlight of your attention onto the sensation but don't allow yourself to get caught up in it. Imagine that on the in-breath you are gently breathing air into the location where the sensation is strongest, then expelling it on the out-breath. You may notice that when you explore the sensation with friendly curiosity—not trying to change it in any way, neither clinging to it nor repressing it—the feeling will start to fade of its own accord. When it has gone, return your full attention to the breath.

Mindfulness instructors sometimes talk about "surfing" the wave of an unpleasant sensation such as pain, anxiety, or craving. Instead of allowing yourself to be overwhelmed by the wave of feeling, you get up on your mental surfboard and ride it. You experience it fully, but your mind remains detached, dignified, and balanced. Knowing that the power of even the most fearsome wave eventually dissipates, you ride it out.

If a thought, emotion, or feeling becomes too strong or intrusive, you can always use the breath as a calm refuge, returning your whole attention to the breathing sensations at your nostrils. Similarly, if you feel you can't cope with a pain such as stiffness in your legs, neck, or back, shift your posture accordingly. But make your intention to move a mindful choice rather than a reflex, and make the movement itself slow and deliberate.

CHAPTER FIVE

THE MAN WHO
DISAPPEARED

The putting away of that pride which comes from the
thought "I am!" this truly is the highest happiness.
　　　　　　　　　　—The Mahāvagga, chapter 1, part 3,
　　　　　　　　　　　　translated by T. W. Rhys Davids

On the day he arrived after his long walk from Uruvelā (later
known as Bodh Gaya), Siddhārtha arranged for their encampment
to be relocated within the shelter of a monumental banyan tree.
For the next three days—protected from the sun and the monsoon
rains, adequately fed and carefully instructed—his five disciples re-
gained their strength and morale. On the fourth morning, when
they returned from the town, Siddhārtha judged them ready to
learn of the final insight they must realize in order to attain enlightenment.[1, 2]

"The body is not the Self," he begins. If this body were the Self—the eternal essence of a person, their soul—would it be subject to so much disease and change? Would it be beyond our control, perishable? Would it be the cause of so much suffering and unsatisfactoriness, just like everything else that is impermanent by nature?

"No, venerable sir," they reply. "It would not." Whatever form the body has taken, or takes, or will take, Siddhārtha tells them, you surely have to conclude: "This is not mine, this is not I, this is not my Self."

He then asks them to scrutinize each of the four remaining components of a sentient being in turn and judge for themselves whether it is permanent or impermanent. Feeling, perception, thoughts, consciousness . . . ? "Impermanent," they respond every time. So, if there is no refuge of stability to be found in any of the five components, he says, where could the supposedly eternal, unchanging Self possibly reside within a human being? Everywhere one looks, whether in the body or the mind, one sees only change and instability. The Self is nowhere to be found.

For a thousand years, the central tenet of the Vedic religion of ancient India had remained unchallenged: inside every human dwells Atman, an essence that is his or her true Self. This "inner controller" is indistinguishable from the ultimate reality, Brahman, which is the Supreme Soul and ground of all existence. In the deepest, most ecstatic states of meditation, a yogin was said to discard his mundane, everyday self and realize his true Self, becoming one with Brahman and every other sentient being on earth and in the heavenly realms. The first recorded use of the Sanskrit word *Atman* was in the Rig Veda,[3] a collection of sacred hymns that may date as far back as 1500 BCE. The Brihadaranyaka Upanishad, a treatise on Atman composed around 700 BCE, neatly summarizes the concept:[4]

> *He is never seen, but is the Witness; He is never heard, but is the Hearer; He is never thought, but is the Thinker; He is never known, but is the Knower. There is no other witness but Him, no other hearer but Him, no other thinker but Him, no other knower but Him. He is the Internal Ruler, your own immortal Self.*

The Vedic priests taught that if a human being managed through the disciplines of Yoga to strip away their selfish, egotistical urges—their crude, lowercase "self"—they would be left with pure awareness, the witness of their every thought and sensation, their true Self. Crucially, this entity was eternal, unchanging, and indistinguishable from the Supreme Soul. Throughout human history, spiritual seekers of all kinds have reached similar conclusions about the importance of apprehending Self. In ancient Greece, the maxim "Know thyself" was inscribed in stone at the Temple of Apollo in Delphi;[5] the God of the Old Testament, speaking to Moses out of the burning bush, declared his name to be "I am that I am";[6] Christian Gnostics believed that only by investigating the Self could one find God;[7] and in Sufism, the "secret soul" or *ruh sirr* was conceived as a direct connection to the divine.[8] To comprehend the Self was to understand the supreme divinity, the ultimate "I am."

In ancient India there were yogins who were convinced they could discern their very Self during the deepest stages of meditation. The author of the Bhagavad Gita wrote that when meditation had been mastered, the mind was "unwavering like the flame of a lamp in a windless place. In the still mind, in the depths of meditation, the Self reveals itself. Beholding the Self by means of the Self, an aspirant knows the joy and peace of complete fulfillment." With persistence, the writer says, the yogin could achieve unity with the source of all existence: "Wherever the mind wanders, restless and diffuse in its search for satisfaction without, lead it within; train it to rest in the Self. Abiding joy comes to those who still the mind. Freeing themselves from the taint of self-will, with their consciousness unified, they become one with Brahman." [9]

Accordingly, the brahmins envisaged the Self as a spiritual being—living within the corporeal being—that was indivisible, unchanging, and beyond suffering. But when Siddhārtha investigated the five components or "aggregates" of human existence one by one—our physical form, feelings, perceptions, volitions, and consciousness (in other words, the entirety of a person)—he discovered that every single one was ultimately beyond our control, impermanent and therefore unsatisfactory. Each was bound up with suffering. You could believe there was something else—an ethereal spirit or soul disconnected from every other known mental and physical phenomenon—but that would be purely a matter of faith. In Siddhārtha's opinion, it was a delusion. He had spent six years refining the skills needed to investigate his own mind, matching or exceeding those of anyone else alive. Through mindfulness meditation he had searched long and hard but encountered nothing that met the minimum requirements to justify the appellation Atman, his eternal, unchanging Self. His subsequent teaching of "nonself" was not a trivial claim, because without access to a Self, how could one hope to commune with the universal "I am," Brahman? When you ceased to believe in the Self, the sacred bridge that linked humanity to the divine vanished like a mirage and with it any hope of channeling its power, as the brahmins claimed to do through their fiery sacrifices.

Intimations that the soul does not exist are not unique to Buddhism, though they are nearly always considered incompatible with religious faith. This would prove treacherous terrain for the anonymous Christian author of *The Cloud of Unknowing*, who believed that during meditation one could reach out and become "perfectly united to God" while still here on earth. He wrote that the sinful Self was a source of great sorrow, "a foul, stinking lump" that must always be "hated and despised."[10] Wouldn't it be better to deny its very existence? The writer seems to have been on the brink of reaching the same conclusion as Siddhārtha—before recoiling from it in horror: "Yet in all this sorrow the soul does not desire not to exist, for that would be diabolical madness and contempt for God; but it is well pleased to exist and offers heartfelt thanks to God for the excellent gift of existence, even though it desires unceasingly to lack for knowledge and feeling of its existence."

Aside from this theological trepidation, there was a psychological barrier to accepting the concept of "nonself." The Buddha taught that while each of us is simply an aggregation of elements in a state of constant flux, our clinging or attachment to them— our *identification* with them—gives rise to suffering and the delusion of selfhood. According to this view, a human being is a process without any fixed center. The twentieth-century translator and scholar Eknath Easwaran, an expert in the philosophies of ancient India, compared the five aggregates that Buddhists believe comprise a person to the brightly colored spices one buys in the marketplace before taking them home to grind and combine for that night's dinner.[11] Thus, from moment to moment, each of us is no more than a unique blend of spices, a homemade garam masala. Surely, you may be thinking, there must be more to "me" than that? To be human is to have a sense of owning one's sensations, thoughts, feelings, and urges. We intuitively believe there is a thinker behind our thoughts, a hearer behind what we hear, a viewer behind what we see. We have the sense of being an operator inside our own skulls, looking out at the world through our eyes and calling the shots.

If that were really true, a human being would be little different from the jewelry shop owner Rosenberg in the movie *Men in Black*, who carries around an extraordinary secret within his head. When this gray-haired, bespectacled jeweler winds up in a hospital mortuary after being killed in a fight, Agents J and K quickly discover that he was no more than an elaborate machine built to look like a human, operated by a diminutive alien, a "homunculus" who sits inside his cockpit-like skull, pulling levers with his tiny hands and pushing little pedals with his feet. You might wonder, quite reasonably, whether there could be another homunculus operating the alien? Is there an even smaller creature sitting inside *his* head, and so on, ad infinitum? The trouble with the idea of an "inner controller" who runs our lives through the exercise of free will is that it presupposes there can be actions without prior causes, which is impossible. And yet that is what we continually tell ourselves must be happening.

In the eighteenth century, the philosopher David Hume reached the same conclusion that Siddhārtha Gautama had in the fifth century BCE: [12]

If any impression gives rise to the idea of self, that impression must continue invariably the same, through the

*whole course of our lives; since self is supposed to exist
after that manner. But there is no impression constant
and invariable. Pain and pleasure, grief and joy, passions
and sensations succeed each other, and never all exist at
the same time. It cannot, therefore, be from any of these
impressions, or from any other, that the idea of self is
derived; and consequently there is no such idea. . . . For
my part, when I enter most intimately into what I call
myself, I always stumble on some particular perception or
other, of heat or cold, light or shade, love or hatred, pain
or pleasure. I never can catch myself at any time without
a perception, and never can observe any thing but the
perception. . . . I may venture to affirm of the rest of mankind, that they are nothing but a bundle or collection of
different perceptions, which succeed each other with an
inconceivable rapidity, and are in a perpetual flux and
movement.*

To assert that the Self does not exist was certainly considered
revolutionary in the Buddha's day, and it will still strike many
modern readers as peculiar. But it is important to understand that
Siddhārtha was not denying the *experience* of selfhood, which is perfectly real. It is our stream of consciousness. But the thoughts, emotions, and memories that comprise it come and go, our bodies and
minds change. Even consciousness waxes and wanes throughout
the day and while we are asleep. Our components are in a constant
state of flux. "The mind is a kind of theatre," said Hume, "where
several perceptions successively make their appearance; pass, re-
pass, glide away, and mingle in an infinite variety of postures and
situations." In Buddhist terminology, the aggregates (form, feelings,

perceptions, volitions, and consciousness) are "conditioned"—they are being carried along in a stream of causes and effects. The goal of mindfulness is to resist our natural tendency to cling to them, and instead step out of the stream and simply watch them go by. Only then can we see through the illusion of selfhood created when the mind identifies with the flotsam of consciousness. What you're left with isn't nothing—annihilation—but *knowing*. The Pāli word *Buddha* can be translated as "one who knows." This knowing is of course only possible as a result of consciousness, which remains something of a scientific mystery. Crucially, however, its operation is not dependent on our having an internal controller, a soul, or Self.

For Buddhists, to internalize the teaching of nonself (*anatta* in Pāli) has a remarkable consequence. It starts to break down distinctions between "you and me," or "us and them," because when the proud, egotistical "self" is taken out of the picture, these differences blur and the sense of alienation we might once have felt begins to dissolve. This change in perspective also loosens the grip of jealousy and possessiveness, rendering the difference between "yours" and "mine" unimportant.

Like many people, when I first came across the teaching of nonself, I concluded that Buddhists are nihilists bent on becoming shaved-headed, impassive automata. But the abbot of Amaravati Monastery, Ajahn Amaro, quickly put me straight. "It's about recognizing that the boundaries of 'I' and 'me' and 'mine' are convenient fictions," he told me during our interview. We were sitting in the study of his bungalow-like kuti overlooking a tiny walled garden. I was surprised to see the unmistakable face of the Dalai Lama beaming down at us. It felt like finding a painting of the pope hanging on the Archbishop of Canterbury's wall, but was further evidence that Buddhists do things a little differently. The abbot was sitting at his desk, on which I had placed my recorder for the interview. He was in full flow, only occasionally pausing to take a sip from a mug. "You call this 'your' recording device," he said, pointing at the machine, "but if it falls out of your pocket and you leave it on the tube, it's not 'yours' anymore. The idea or the memory lingers but the object is no longer yours." I reflexively checked whether its red light was lit, indicating that it was still recording all this. "Anatta [nonself] is a recognition that this being cannot be divorced from everything else. We're constantly breathing air in and breathing air out, eating food, excreting, living in interrelationship with the whole world, both material and mental, biological and physical. There's an unrelenting exchange, an interrelationship. But it's not like saying 'I don't exist—I'm a Buddhist, I'm supposed to believe I don't exist.' That's a common misunderstanding. It's a bit like saying . . ." He swept his arm across the desk to illustrate the point and knocked his mug flying, sending a shower of brown fizzy liquid everywhere. "Woops!"

I rescued my recorder from the puddle in which it was now sitting. "Do you want to test it to see if it's still going?" he said, clearly concerned. "It's still going," I replied optimistically, checking the light, "it was just a few drops." The monk went into the adjoining room and brought back some paper towels. "So, technically that was *my* drink," he said jokingly. "Now it's not anymore! We're going to get wet feet . . ." He got down on his knees to swab the floor and from this position continued where he had left off. "It's also kind of interesting, someone was telling me last year how Freud coined the word *ego* without the idea of it being some sort of fixed, separate entity or quality, he was merely using the Latin word as a way to talk about the *feeling* of 'I am.' A shorthand for the

'I' feeling, which is very comparable to the Buddhist approach." He went to the other room for more paper towels without breaking his flow. "As part of our experience, that word is applicable—'I am speaking,' 'I am here,' 'I am feeling'—but when that word is imputed with absolute reality, a distinctness that's making it separate from the rest of the world, then there's trouble."

Buddhists believe that to realize this truth—that you are not your Self—is immensely liberating, and so by the time the Buddha had finished speaking after delivering this second lesson to the five monks in the deer park, the scriptures say there were now six *arahants* (fully enlightened beings) in the world.[13] With the teaching of nonself, alongside the teachings about suffering and impermanence, they had realized nibbāna: they were free from all worldly attachments and the taints of craving, aversion, and delusion. The Pāli word *nibbāna* means literally to be "snuffed out," but it was their egos that had been extinguished, not their personalities.

Spiritual enlightenment is often portrayed in these Buddhist stories as coming suddenly, like a bolt of lightning. The progress of scientific enlightenment over the past five hundred years has been rather more sedate but no less dramatic, dealing a series of humiliating blows to the human ego. In the sixteenth and seventeenth centuries, Copernicus, Kepler, and Galileo managed to figure out that Earth is not the center of the universe. In the nineteenth century, Darwin proposed that all creatures had evolved through natural selection from a simple common ancestor, and that humans were just another kind of ape. In the twentieth century we learned that our sun is an insignificant star among the hundred billion stars in the Milky Way, which is just one of around 170 billion galaxies in the known universe. In the early years of the twenty-first century, astronomers have concluded that many, if not most, of these other galaxies contain habitable planets, some of which may be home to more advanced civilizations than our own. To cap it all, we have to listen to scientists and philosophers questioning whether we humans have free will, let alone a Self.

In his book *Chance and Necessity*, French molecular biologist and Nobel Prize winner Jacques Monod reflected upon our species' perennial struggle to maintain its special status in the universe, despite growing evidence to the contrary: "We would like to think ourselves necessary, inevitable, ordained from all eternity. All religions, nearly all philosophies, and even a part of science testify to the unwearying, heroic effort of mankind desperately denying its own contingency."[14] Perhaps it gives us a sense of security, a sense of belonging, in much the same way that a child needs a secure, nurturing environment to develop a strong sense of self while growing up. Ajahn Amaro uses the metaphor of an embryonic bird inside its protective shell to make an important distinction between this vital stage in our early lives and the stage when we are sufficiently mature to break out of our shells and confront the real world.

At first it will seem deeply counterintuitive, but it gets easier. The notion that inside each human head is a pilot—an unchanging, indivisible Self like the alien sitting in Rosenberg's skull—is very convincing until you start to notice how that experience of selfhood changes not just over a lifespan but from moment to moment, from hour to hour, and from day to day. Notice how the feeling of being "you" changes according to whether you have drunk your morning coffee, how well you slept, how full your stomach is, whether you are feeling distracted, irritated, bored, worried, or elated. If you are anything like me, the experience of selfhood never stays the same from one minute to the next. Now recall how it felt to be you at the last office Christmas party with your colleagues; when you were sitting eating Christmas dinner with your family; a week later on New Year's Eve at a party with friends. Did you feel and behave like the same person on each of these occasions? Which one was the *real* you? Perhaps you took a brief holiday before returning to work in the new year. What was it like to be you relaxing on a beach with just the sound of the waves, perhaps reading a novel to "take you out of yourself"?

Of course, each of us has distinctive personality traits, but even these vary according to the circumstances. Our place on the spectrum from shy introversion to uninhibited extroversion, for example, will vary according to the setting, the people we're around—how well we know them and our relationship to them—whether we have had a drink, our alertness, how happy we feel at that particular moment. We hear expressions like "She was not herself" all the time to justify some indiscretion, but when are we ever our true selves?

The last refuge of the Self, perhaps, is "physical continuity." Despite the body's mercurial nature, it feels like a badge of identity we have carried since the time of our earliest childhood memories. A thought experiment dreamed up in the 1980s by British philosopher Derek Parfit illustrates how important—yet deceiving—this sense of physical continuity is to us.[15] He invites us to imagine a future in which the limitations of conventional space travel—of transporting the frail human body to another planet at relatively slow speeds—have been solved by beaming radio waves encoding all the data needed to assemble the passenger to their chosen destination. You step into a machine resembling a photo booth, called a teletransporter, which logs every atom in your body then sends the information at the speed of light to a replicator on Mars, say.

This rebuilds your body atom by atom using local stocks of carbon, oxygen, hydrogen, and so on. Unfortunately, the high energies needed to scan your body with the required precision vaporize it— but that's okay because the replicator on Mars faithfully reproduces the structure of your brain nerve by nerve, synapse by synapse. You step into the teletransporter, press the green button, and an instant later materialize on Mars and can continue your existence where you left off. The person who steps out of the machine at the other end not only looks just like you, but etched into his or her brain are all your personality traits and memories, right down to the memory of eating breakfast that morning and your last thought before you pressed the green button.

If you are a fan of *Star Trek*, you may be perfectly happy to use this new mode of space travel, since this is more or less what the USS *Enterprise*'s transporter does when it beams its crew down to alien planets and back up again. But now Parfit asks us to imagine that a few years after you first use the teletransporter comes the announcement that it has been upgraded in such a way that your original body can be scanned without destroying it. You decide to give it a go. You pay the fare, step into the booth, and press the button. Nothing seems to happen, apart from a slight tingling sensation, but you wait patiently and sure enough, forty-five minutes later, an image of your new self pops up on the video link and you spend the next few minutes having a surreal conversation with yourself on Mars. Then comes some bad news. A technician cheerfully informs you that there have been some teething problems with the upgraded teletransporter. The scanning process has irreparably damaged your internal organs, so whereas your replica on Mars is absolutely fine and will carry on your life where you left off, this body here on Earth will die within a few hours. Would you care to accompany her to the mortuary?

Now how do you feel? There is no difference in outcome between this scenario and what happened in the old scanner—there will still be one surviving "you"—but now it somehow feels as though it's the *real* you facing the horror of imminent annihilation. Parfit nevertheless uses this thought experiment to argue that the only criterion that can rationally be used to judge whether a person has survived is not the physical continuity of a body but "psychological continuity"—having the same memories and personality traits as the most recent version of yourself. Buddhists have formulated what is essentially a theory of individual psychological continuity. Called "dependent origination," it envisages a causal chain of events whereby someone's motives and actions and the associated kamma determine what happens to them as their life unfolds (see Figure 8, page 282). The chain is pictured as extending beyond death, affecting rebirth into a future life and a different physical body in a way not dissimilar from an interplanetary teletransporter. Many Buddhists believe that the kamma resulting from their attitudes and behaviors in this life will determine their fate in this future existence, and some even claim to remember past lives.

The Buddhist concept of rebirth is controversial (I'll return to it in the final chapter), but few would quibble with the idea that what we do today can affect our lives further down the line. If you take up smoking, for example, your older self will have to face the consequences. Given the opportunity, they'd likely chide their younger self. The trouble is that young people can barely conceive that they will ever be old, leading them to take chances with their health or safety. The younger and older selves don't see eye to eye and the supposedly indivisible Self is divided against itself across time. But different parts of the brain can also come into conflict about lifestyle choices, even to the extent where there appear to be two or more "Selves" battling for control. One of its hemispheres might be in favor of smoking while the other is vehemently against it. Our brains resolve such conflicts without our noticing, yet they became very apparent, for instance, in a woman who developed "alien hand syndrome" after a stroke damaged her corpus callosum, the bundle of fibers that links the right and left hemispheres of the brain and normally coordinates their activity. Researchers observed her putting a cigarette between her lips with her right hand (controlled by the left hemisphere), only for her left hand (controlled by the right hemisphere) to take it out and throw it away before the right hand could light it. "I guess 'he' doesn't want me to smoke that cigarette," the patient observed wryly.[16] People with alien hand syndrome swear they don't have any influence over what the rogue hand gets up to: it's as if another, independent Self is in control of it.

Other "split brain" effects can be caused by an operation to prevent epileptic seizures that was pioneered in the 1960s by neurobiologists Roger Sperry and Ronald Meyers. It involves severing the corpus callosum and works like a firebreak in a blazing forest, stopping the spread of uncontrolled electrical activity from one side of the brain to the other. But it also curtails normal communication between the two hemispheres. People who have had the operation can appear perfectly normal, showing no noticeable change in temperament, personality, or intelligence, and no difficulty in social situations, but on closer examination, something extraordinary becomes apparent. Thanks to the division of labor between the two halves of a brain, researchers have been able to investigate the consequences of severing the communication link. The left hemisphere processes sensory inputs from the right side of the body—receiving visual information from the right eye, for example—and controls voluntary movements on that side, whereas the right hemisphere is in charge of the other side of the body. In addition, the left hemisphere specializes in reading and speech, whereas the right hemisphere has only rudimentary language abilities. For instance, when researchers show a picture of a spoon to a patient's left eye only and ask her what she can see, she will deny being able to see anything at all. But when asked to reach behind a partition with her left hand and feel around for the thing she "did not see" among a variety of objects, she will pick up a spoon. When asked to say what the object is in her hand, she will be unable to name or describe it.[17]

Other studies of patients whose corpus callosum has been severed demonstrate that the two hemispheres can learn things independently of each other, and they have separate memories and competing motivations. One side of the brain can know something that the other is completely unaware of, and one side can be opposed to the actions of the other. To all intents and purposes, each hemisphere is now hosting its own stream of consciousness: there are now two Selves operating out of the same head. It's as if Agents J and K were to open up Rosenberg's skull and find a pair of aliens sitting side by side.

Split-brain patients dispel any notion that inside each of us is an unchanging, indivisible Self—an inner controller sitting in splendid isolation pulling our body's levers. The two hemispheres are one source of division; another is that between the primitive, emotional parts of the brain and the more recently evolved regions:

there is a constant tug of war between the fearful, angry animal inside all of us and the more intellectual being who worries about long-term consequences. Our brains also contain opposing networks of regions that vie with each other to determine whether our minds are focused on an external task or engaged in internally directed thoughts. On the basis of all this evidence, neuroscientist David Eagleman of Baylor College of Medicine in Houston, Texas, has proposed that we should view the brain not as a single unified system but as a team of rivals squabbling among themselves. Often we are not even aware of these rivalries until a stroke or brain surgery exposes them.[18] I'm reminded of the ingenious Disney-Pixar animated film *Inside Out*, which depicts a team of five operators at the controls inside a little girl's head—Joy, Sadness, Anger, Fear, and Disgust—who have very different ideas about how she should behave. While our five basic emotions are not generated by distinct parts of the brain operating in isolation, we can recognize how they compete for control over our lives.

Amid all this internal confusion and rivalry, how can we know ourselves at all? How do we remember from day to day what kind of a person we are—proud, sensitive, obstinate, vivacious, shy? Toward the end of my stay at the monastery, I attended a meditation class where a young monk suggested we should ask ourselves "Who am I?" without expecting any answer. It's the kind of mental exercise Buddhists enjoy. When I got home I asked my smartphone the same question. The reply came back: "You're asking me, James?" It then displayed my address and telephone numbers. I rephrased the question. "*What* am I?" Within milliseconds the phone responded: "Human." Is there a part of the brain where the sense of selfhood is generated—not your species, name, or contact details but *what you are like*? In the past, some neuroscientists claimed there was nothing special about the way the brain handled such information, arguing that how we remember things about ourselves is no different from the way we remember things about anything—pop singers, Popsicles, popes—it's just that we're much more familiar with "us" than with anything else in the world. However, recent research has identified a part of the brain that specializes in reporting to consciousness not only the kind of people we are but also how we relate to those around us.

Scientists at Dartmouth College in Hanover, New Hampshire, tracked down this activity of the brain in 2002 by asking people to respond to a series of adjectives describing personality traits such as "dependable," "polite," and "talkative" while they lay in an MRI scanner. They were told to respond in one of three ways to each word: "Does the adjective describe you?" (self); "Does the adjective describe the current US president, George Bush?" (other); and "Is the adjective in uppercase letters?" (case). When they were asked whether the adjective applied to them, it caused increased activity in the volunteers' medial prefrontal cortex, compared with when they were asked "other" or "case" questions.[19] Subsequent research has confirmed this finding and singled out adjacent areas, including the cingulate cortex, which fire up not only during self-reflection but also when we're intuiting what others are thinking.[20] This suggests that the same parts of the brain simulate both the feeling of "self" and how it feels to be somebody else, in other words, empathy. So it makes sense that the more difficulty that people on the autism spectrum have intuiting the thoughts and motivations of others, the less sense of their own self they have. What's more, this diminishing sense of selfhood and worsening social ability seems to correlate with weaker activation in the cingulate cortex of their brains.[21]

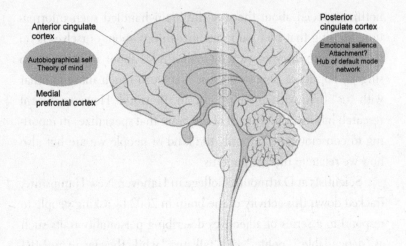

Figure 3. The brain's "Self app." The medial prefrontal cortex, cingulate cortex, and other cortical midline structures are activated not only when we think about ourselves but also when we intuit the intentions and beliefs of other people, a skill known as "mentalizing" or "theory of mind." They draw upon long-term, autobiographical memories encoded and stored in the temporal lobe. Meditation appears to reduce self-referential thinking by decreasing activity in the medial prefrontal cortex and posterior cingulate cortex, a principal node of the default mode network.

Neuroscientists often refer to these neighboring regions as a single unit—the cortical midline structures, or CMS. They are found on the inner surface of each hemisphere, hence the adjective "medial" (see Figure 3, above). If you were to take a human brain and pry apart the two hemispheres to expose their inner faces—like opening up a sandwich to inspect the filling—you would be looking at the CMS of each. In a sense, the brain is an ego sandwich. Evidence is growing that the CMS are core components of the brain's "Self app": they not only access autobiographical memory and the relatively stable aspects of our personality but also somehow "bond" these to important stimuli in our social and physical environment, thus creating a sense of "me" and "others."

This has led some researchers to speculate that the sense of self is simply a by-product of a system that evolved to help us relate to other people. I explore some surprising implications of this in chapter 9, "The Fall," but it's worth noting here that two areas of the CMS that light up when people are asked to reflect about their own personality—the medial prefrontal cortex and posterior cingulate cortex—are proportionally bigger in humans than in any other primate. They are also well connected with each other, leading some neuroscientists to propose that the posterior cingulate cortex interfaces between us and the outside world by gauging the personal and emotional importance of stimuli, while the medial prefrontal cortex is where this information interfaces with consciousness.[22, 23]

I asked Judson Brewer, a psychiatrist and mindfulness researcher at Yale University School of Medicine and the University of Massachusetts Medical School, to explain how the two regions might work together to create this feeling of Self. "One theory is that the medial prefrontal cortex is a hub of the conceptual self—'I'm Jud'—whereas the posterior cingulate cortex is an experiential kind of 'clenching' or 'time stamp' that links that conceptual self with events in the world," he said. "So we feel like 'I'm the one who got angry in this situation. I'm the one doing this.'" Brewer has proposed that the posterior cingulate creates the experience of being "caught up" in our thoughts, feelings, and sensations—which corresponds closely with the Buddhist concept of grasping or attachment identified as the cause of dukkha or suffering—and also that it may play an important role in drug addiction.[24] He has gathered evidence showing that by toning down the activity of the posterior cingulate cortex, mindfulness can reduce nicotine cravings in smokers who want to quit. Clinical trials suggest it may also have a role in treating a wide range of other cravings, whether for food, drugs, or anything else we desire (see chapter 7, "Fire Worshippers").

By contrast, the psychedelic psilocybin—the active component of so-called magic mushrooms—is not physically addictive, though it does have a dramatic effect on the two key nodes of the CMS. An fMRI study in 2012 by researchers at Imperial College London found that it not only quiets the medial prefrontal and posterior cingulate cortices but also uncouples their activity.[25] This may explain the "egoless" state reported by people who take psilocybin, in which the distinction between self and the rest of the world dissolves, evoking the profound spiritual experience of being "at one with the universe." It is as if taking the drug temporarily short-circuits the brain's machinery for generating our sense of "myself" and "everything else." Natural psychedelics have been used in traditional religious rites to channel the divine in much the same way that the Vedic priests in ancient India believed they could unite Atman with Brahman, the Self with the Supreme Soul, through meditation. It has even been speculated that their trance-inducing ritual drink soma was made from a type of psychedelic fungus.[26]

Crucially, the CMS are part of a network of brain regions known as the "default mode network" that become active when we're not focused on a particular task and our minds are wandering.[27] Another key node in this network is the hippocampus, where the brain encodes "episodic" or "autobiographical" memories—recollections of things that have happened to us. By contrast, when we're focused on performing a task that requires our full attention, a constellation of regions known as the task-positive network fires up as activity in the CMS and the mind-wandering network dies down. So it seems the brain has two opposing networks: one for handling tasks that demand focused attention, and another for self-reflection, musing about past experiences, and speculating about the future. While one is activated the other is suppressed, and vice versa. These findings accord well with the pleasurable experience we can sometimes have of "losing ourselves" in a task that absorbs our whole attention, whether it's a video game, motorcycle maintenance, or mathematics. This is the elusive sense of "flow" that musicians and athletes aim for, seeking their best possible performance. While we're "in the zone" and completely focused on a task, any meandering thoughts about ourselves fall away. You may also be familiar with the opposite experience, being so consumed by your thoughts you forget what you're meant to be doing or saying, perhaps becoming clumsy, making silly mistakes in your work, or losing the thread of a conversation. "Sorry, I was miles away," you say.

The objective of mindfulness meditation is to notice whenever your attention starts to get lost in this self-referential maze of thoughts and bring it back to the breath or whatever sensation has been chosen as the meditative focus. Buddhists believe this trains the mind to appreciate that our sense of self is not a solid, unchanging thing but is created when we latch on to whatever is bobbing along in our "stream of consciousness"—a term coined by nineteenth-century psychologist William James from a literal translation of the Pāli word *viññāṇa-sota*.[28] The idea is that with practice one learns how to let go of passing thoughts, memories, and emotions. Remarkably, the neurological signature of this shift in perspective can be read in the brain after only eight weeks of mindfulness instruction. When people who have taken the course are asked to focus their minds on moment-to-moment experiences as they lie in an MRI scanner, activity in their medial prefrontal cortex—one of the two key components of the brain's "Self app"—declines significantly compared with control participants who haven't received any training.[29] Interestingly, at the same time there is greater activity in brain regions involved in body awareness and processing sensory stimuli, which as we saw in the previous chapter are also associated with the pain-relieving effects of meditation. What seems to be happening with mindfulness is that it shifts our focus away from self-referential thinking toward a more detached, moment-to-moment focus on bodily sensations and our environment.

With dedicated practice, this altered perspective becomes more ingrained, so that the characteristic changes in brain activity are not restricted to periods of meditation but become a permanent feature of waking life. In 2011, Brewer and his colleagues at Yale University found that the medial prefrontal and posterior cingulate cortices of people who had been practicing for more than ten years were less active while they meditated in an MRI scanner compared with control subjects.[30] Even when not meditating, there were distinctive differences between the two groups. In meditators, activity in parts of their brain involved in meta-awareness (the ability to think about thinking) and cognitive control rose and fell in conjunction with that in the posterior cingulate cortex, suggesting that its activity was being regulated in some way.

The Buddha's teaching of nonself goes to the heart of what it means to attain enlightenment or nibbāna, because *understanding* the concept is clearly not enough. To fully *realize* it and become free of the habitual "I, me, mine" of human existence requires many years of training the mind not to identify personally with every passing desire and emotion. It involves slowly erasing old mental habits and inscribing new ones into the hardware of the brain. Trainee meditators learn to notice when they are grasping at thoughts, emotions, and body sensations and then let go, bringing their attention back to the present moment. Instead of thinking, for example, "I am angry" when this emotion arises, they are encouraged to say silently, "there is anger." In this way they start to develop a more impersonal, third-person perspective on their stream of consciousness. Far from being a mere philosophical, "spiritual" exercise, this decentering of the mind has a measurable effect on well-being, as we shall discover in the next chapter. Mindfulness-based therapies are proving their worth for reducing stress and anxiety and for preventing relapse in those prone to depression.

Buddhists believe that our deeply ingrained sense of having an independent, unchanging Self that miraculously stands aloof from everything and everyone else is not only illusory but opens us up to all manner of suffering. A great twentieth-century physicist who famously tore up the rules of space and time reached a similar conclusion. In 1950 he wrote a letter to Robert E. Marcus, political director of the World Jewish Congress, whose son had recently died from polio. It was not a conventional message of condolence by any means: [31]

Dear Dr. Marcus

A human being is part of the whole, called by us "Universe," a part limited in time and space. He experiences himself, his thoughts and feelings as something separated from the rest—a kind of optical delusion of his consciousness. The striving to free oneself from this delusion is the one issue of true religion. Not to nourish

the delusion but to try to overcome it is the way to reach
the attainable measure of peace of mind.

> With my best wishes,
> sincerely yours,
> Albert Einstein.

This is easier said than done, of course. But the evidence gathered by Brewer at Yale and others suggests that meditation might
help us attain this happy state.

It's tempting to imagine what an fMRI analysis of Siddhārtha's
brain would have revealed if scans were made at regular intervals
in the six years following his abandonment of the comfortable
home life, during which he followed a strict regime of meditation.
Scientists surely would have recorded, as they have in the brains of
today's committed meditators, that the parts of his brain involved
in monitoring moment-to-moment experience and exercising cognitive control were becoming more powerful. At the same time,
they would have seen steadily decreasing activity in Siddhārtha's
cortical midline structures, the regions most closely associated
with creating the delusion of having a solid, unchanging Self.

After his enlightenment, Buddhist scriptures tell us that the
man who had once been called Siddhārtha Gautama began to
refer to himself in the third person, as the Tathāgata—which literally translated means "thus gone."[32] He no longer clung to the
impermanent constituents of his being and the old delusions of
selfhood. Buddhists believe that by extinguishing the Self, he extinguished suffering. To all intents and purposes, the man he had
once been had vanished.

CHAPTER SIX

GOLDEN SLIPPERS

Hard it is to train the mind, which goes where it likes and does what it wants, but a trained mind brings health and happiness.

—The Dhammapada (translated by
Eknath Easwaran), verses 35–36

It was around four o'clock in the morning when the millionaire's son materialized from the darkness like a sleepwalker, wearing nothing but a silk dressing gown and a pair of slippers trimmed with gold thread. He was in a bad way, groaning and talking to himself about danger and distress. His golden slippers were caked with mud from his walk across the rain-sodden park. What had happened, it later transpired, was that he had awoken in the small hours of the morning in his mansion in Benares (Varanasi) and, unable to get back to sleep, a steadily worsening sense of dread had risen up inside him. He got out of bed, pulled on his dressing gown and slippers, and wandered out into the night. His name was Yasa and he was the son of a wealthy merchant.[1, 2] Everyone in the encampment beneath the banyan tree was slumbering peacefully when he appeared, apart from Siddhārtha, who had been practicing walking meditation when he saw the distressed stranger. He approached the young man and reassured him that he had reached a place of safety. Then he encouraged him to sit down for a while and talk things through. Would he perhaps like to learn the truth about suffering, its cause, its ending, and the noble path leading toward its ending? The man cast away his muddy gold slippers and sat down to listen to what the Buddha had to say.

For many people, to be left alone with their thoughts is a kind of torture. Insomniacs who have suffered agonies as they lie awake night after night soon learn that it is far better to get up and do something, *anything*, rather than thrash about with only their restless mind for company in the vain hope of eventually falling asleep. Negative emotions such as guilt, self-doubt, and anxiety run amok at night, when one is alone in the dark. Daylight, with its promise of mundane tasks and social interaction, usually sends these monsters of our imagination scurrying back to their caves, but they can reemerge whenever there are no external distractions to occupy the mind. Some people will go to extraordinary lengths to prevent this from happening, as a series of experiments by psychologists at Harvard and the University of Virginia demonstrated in 2014.[3]

College students were instructed to sit by themselves for up to fifteen minutes in a sparsely furnished, unadorned room and "entertain themselves with their thoughts." They were allowed to think about whatever they liked, the only rules being that they should remain in their seat and stay awake. Before they entered the room they were obliged to surrender any means of distraction they had about their person, such as cell phones, books, or writing materials. Afterward, they were asked to rate the experience on various scales. Unsurprisingly, a majority reported that they found it difficult to concentrate and their minds had wandered, with around half saying they didn't enjoy the experience.

A subsequent experiment, however, revealed that many found being left alone in an empty room with nothing to occupy their minds so unpleasant (this is, after all, what makes solitary confinement such a harsh punishment in prisons) that they would rather give themselves electric shocks. In the first part of this experiment, the volunteers were asked to rate the unpleasantness of a shock delivered via electrodes attached to their ankle and say whether they would pay a small amount of money to avoid having to experience it again. In the second part, during which they were left alone with their thoughts for fifteen minutes, they were presented with the opportunity to zap themselves once again. Amazingly, among those who had said they would pay to avoid a repeat experience, 67 percent of the men (12 out of 18) and 25 percent of the women (6 out of 24) opted to shock themselves at least once. One of the women gave herself nine electric shocks. One of the men subjected himself to no fewer than 190 shocks, though he was considered exceptional—a statistical "outlier"—and his results were excluded from the final analysis.

In their report for the journal *Science*, the researchers write, "What is striking is that simply being alone with their own thoughts for 15 minutes was apparently so aversive that it drove many participants to self-administer an electric shock that they had earlier said they would pay to avoid." This goes a long way toward explaining why many people initially find it so hard to meditate, because to sit quietly with your eyes closed is to invite the mind to wander here, there, and everywhere. In a sense, that is the whole point: we are simply learning to notice when this has happened. So the frustrating realization that your thoughts have been straying—yet again—is a sign of progress rather than failure. Only by noticing the way thoughts ricochet about inside our heads like ball bearings in a pinball machine can we learn to observe them dispassionately and simply let them come to rest, resisting the urge to pull back the mental plunger and fire off more of them. One of the benefits of meditation is that one develops the ability to quiet the mind at will. "Without such training," the psychologists conclude drily in their paper, "people prefer doing to thinking, even if what they are doing is so unpleasant they would normally pay to avoid it. The untutored mind does not like to be alone with itself."

Wandering thoughts, both the pleasant ones and the unpleasant, are a product of the brain's "default mode network," which goes to work whenever our minds are not focused on performing an external task. Researchers first identified the network after noticing that certain brain regions became active during fMRI scans when volunteers were instructed to "lie still and don't do anything in particular."[4] You will remember from the previous chapter that the cortical midline structures of the brain—its "Self app"—are an integral part of this network, which explains why our meandering minds are so self-obsessed. The default mode network allows us to muse about our past and imagine the future. It can rewind to the scenes of events we have lived through and fast-forward to imagine things that haven't yet happened. To do this, it draws on autobiographical memories stored in the medial temporal lobe, in particular the hippocampus—an integral part of the default network. It's not all "me, me, me," though; another important function of the network is to figure out the perspectives of others. What these functions have in common is simulation. Essentially, the network's nodes operate like the hardware of a simulator that draws on past experience to construct our sense of self, conceive the perspectives of other people (which psychologists call "theory of mind"), and conjure up future scenarios (known as "mental time travel"). These capabilities make the network indispensable for social functioning, imagination, creativity, and planning.[5]

We pay a high price for this ingenious neural machinery, though, because the default mode network is responsible for mind-wandering. "Experience sampling"—which involves asking people about their mood and thoughts at random moments throughout the day—suggests that our minds wander from what we're actually doing an amazing 30 percent to 50 percent of the time that we're awake, and that this is often associated with feelings of unhappiness.[6-8] According to Harvard psychologists Matthew Killingsworth and Daniel Gilbert, who created an iPhone app, Rate Your Happiness, to gather some of this data, fluctuations in happiness depend more on what we're thinking than what we're doing. Crucially, the results suggest that mind-wandering is the *cause* rather than the consequence of negative emotions. As the opening verse of the Dhammapada expresses it, "Our life is shaped by our mind; we become what we think. Suffering follows an evil thought as the wheels of a cart follow the oxen that draw it."[9] Less poetically, the psychologists concluded that "the ability to think about what is not happening is a cognitive achievement that comes at an emotional cost."

So, while the default network is essential for virtual time travel and intuiting others' desires, beliefs, and intentions, we must temporarily suppress its activity in order to focus on a mentally demanding task. Evidence is growing that people who are going through an episode of major depression have trouble doing this effectively. Brain-imaging research suggests that they have abnormalities in regions involved in generating and regulating emotions, including the hippocampus, amygdala, and medial prefrontal cortex[10]—all components of the extended default network (they routinely become active when we are not focused on an external task)—and the anterior cingulate cortex, or ACC, which detects errors while we're performing a task and monitors conflicts between different tasks. Compared with control subjects who are not depressed, patients seem to have greater default network activity in response to negative images, for example, and have difficulty restoring this activity to more normal levels when they are asked to reappraise the pictures in a more positive light.[11]

Rumination—a repetitive form of self-referential thinking—is a potent risk factor for clinical depression. In people fortunate enough never to have experienced this debilitating condition, the downsides of a wandering mind may be no more serious than occasional clumsiness, inattention to detail, and momentary unhappiness, but in someone vulnerable to depression, and particularly when they are under a lot of stress, fleeting moments of low mood can become drawn out by rumination and tip over into a full-blown depressive episode. Among individuals who are prone to the condition, research suggests that those who ruminate more tend to have more severe and more prolonged episodes.[12] Overactivity in the frontmost or anterior cortical midline structures (CMS)—particularly the medial prefrontal cortex—appears to play a crucial role in rumination and depression.[13] This makes sense because, as we've seen, this area helps to generate our sense of selfhood and how that relates to the outside world. People who are vulnerable to depression have difficulty keeping a lid on activity in the anterior CMS, either because it is overstimulated by the limbic system—the brain's emotion engine—or as a result of the failure of other parts of the prefrontal cortex involved in emotion regulation to tamp down its activity.

The opening of that last sentence should perhaps have read "people who are *most* vulnerable to depression," because we are all at risk to a greater or lesser extent simply as a result of the way the human brain is wired. Each of us has a default network with a built-in "Self app" that becomes active whenever our conscious attention is not needed to perform a task. Imagine you had an app on your smartphone that automatically fired up whenever you weren't using the phone for anything else, posting repetitive messages of a personal nature on the screen. How annoying would that be? Unfortunately, that is how the human mind appears to work. Recall that up to 50 percent of the time we're awake, our minds are straying from what we're actually doing. This is possible only because much of our waking lives is spent performing tasks that we have carried out so many times before—such as preparing breakfast and eating it, shaving or applying makeup, taking a familiar route to work—that we can safely operate on autopilot without paying conscious attention to what we are doing and what is going on around us. This gives the mind free rein to wander pretty much anywhere it likes, and many of these thoughts will inevitably revolve around our selves and will be negative. When life becomes particularly challenging, for example as a result of problems at work, relationship difficulties, or physical illness, there will be a tendency to ruminate and the potential for persistently low mood and perhaps even depression.

In light of these findings, perhaps it's little wonder that in 2013 the World Health Organization estimated that around 7 percent of the world's population (404 million people) suffer from clinical depression and 4 percent (272 million people) have an anxiety disorder. According to the WHO, depression and anxiety are the leading causes of disability worldwide.[14] By 2030, depression may account for the largest global "disease burden" (a statistic that combines years of life lost due to premature death and years of life lost due to ill health), ahead of common illnesses such as diabetes, heart disease, and infections such as HIV/AIDS.[15] Surveys by the WHO suggest it has a more disabling effect on patients than common physical conditions such as arthritis, angina, and asthma,[16] and one study found that it reduced life expectancy as much as smoking.[17] The toll of suffering for patients and their families is shocking and is reflected perhaps most starkly in the statistics for suicide. According to the latest available figures, around 12 people in every 100,000 took their own lives in the United Kingdom in 2013.[18] In the United States, the rate was 13 suicides for every 100,000 people in 2012.[19]

Over the past two decades, the SSRI class of antidepressants, including Prozac (fluoxetine) and Paxil (paroxetine), and a form of talking therapy known as cognitive behavioral therapy (CBT) have proved their worth for treating persistent depression.[20] Without continuing treatment after recovery, however, there is a 50 to 80 percent risk that the depression will come back.[21] What seems to happen is that negative thinking patterns become ingrained during earlier episodes of the illness, so that even a slight mood disturbance can trigger negative thoughts such as "I'm worthless," "Everything is just too difficult," or "There's nothing I can do to escape my situation." Psychologists call this "cognitive reactivity." Minor bouts of unhappiness, which the person perceives as warning signs of an impending episode, may also lead them to ruminate about their problems in the mistaken belief that this self-analysis will help, when in fact rumination simply prolongs and deepens the initial mood disturbance.[22] "In people who are vulnerable, small amounts of sadness can trigger large increases in negative thinking," says Zindel Segal, a psychologist at the University of Toronto. So, when faced with stressful life events, factors such as low self-esteem and excessive rumination can trigger a relapse.

In 1991, Segal was given seed funding by the MacArthur Foundation to develop a new kind of CBT specifically designed to prevent this from happening. To develop the program, Segal set up a series of meetings in Cambridge, UK, and Toronto, Ontario, with two other specialists in the field of depression relapse: Mark Williams, a clinical psychologist then working at the University of Wales in Bangor, and John Teasdale, a cognitive psychologist at the University of Cambridge. The thinking at the first meeting was that this would simply be a "maintenance" form of the existing program, with monthly rather than weekly therapy sessions. But over the course of the next couple of meetings the three psychologists became more ambitious. They decided they would create a form of CBT that specifically targeted patients' cognitive reactivity: the negative thinking patterns known to be a risk factor for relapse. "We wanted to design a treatment around those triggers that helped people learn how to become less vulnerable, more resilient," Segal told me. "During those discussions we came up with the idea that these depressive patterns of thought could be targeted by teaching people how to relate to their thinking in different ways." But how?

Quite by chance a psychologist named Marsha Linehan from the University of Washington in Seattle was spending part of her sabbatical leave with Teasdale and Williams at the Applied Psychology Unit at the University of Cambridge.[23] In the late 1980s she had developed a program for individuals with borderline personality disorder and based it on an ancient Buddhist practice called mindfulness. She explained that it helped her patients adopt a more objective, accepting perspective on their experience. Rather than trying to "fix" powerful thoughts and emotions, they were encouraged to step back and observe them. Could it perhaps do the same for people with a history of depression? Years before, in 1984, Teasdale had attended a talk about suffering and the practical benefits of meditation by Ajahn Sumedho, an American Buddhist monk, who was Ajahn Amaro's predecessor as abbot of Amaravati Buddhist Monastery in the UK. Teasdale had been struck by parallels between the Buddhist analysis of suffering and his own research into cognitive reactivity in depressed people—but he couldn't figure out how to apply these ancient insights therapeutically. Now some of the pieces of that puzzle were starting to fall into place. All the same, the three psychologists were wary of combining CBT with a practice that mainstream psychiatrists viewed as flaky, if not unadulterated mumbo jumbo. "To talk to my psychiatry colleagues about meditation was to risk career suicide," said Segal, echoing what several other pioneering mindfulness researchers have told me about their early work. So he phoned the MacArthur Foundation to ask how they felt about their money being spent on developing a meditation-based program. The answer came back straight away that they'd be perfectly happy, as long as the program was validated by clinical trials. Nonetheless—all too aware of their fellow psychologists' skepticism about meditation—Segal and his two colleagues kept their heads down for the next couple of years while they developed their new therapy and put it through its paces. They called it mindfulness-based cognitive therapy (MBCT).

In addition to the techniques developed by Jon Kabat-Zinn in the late 1970s and 1980s for his mindfulness-based stress reduction

(MBSR) program, such as mindful movement, mindful eating, and the body scan (see "Guided Meditation: Body Scan," page 236), MBCT includes education about depression and exercises derived from cognitive therapy that are designed to increase awareness of the links between thinking and feeling. In eight weekly group sessions lasting a couple of hours each, plus homework, participants learn to become more consciously aware of bodily sensations, thoughts, and feelings, and to recognize the early signs of a downturn in their mood. Unlike cognitive therapy, however, MBCT does not attempt to change the *content* of negative thinking. Rather, it encourages people to adjust their relationship to thoughts, feelings, and bodily sensations. It allows them to discover for themselves that these experiences are fleeting events in the mind and the body, and that they can choose not to engage with them. Observing thoughts, emotions, and sensations objectively in a spirit of curiosity and self-compassion helps participants understand that they needn't be defined by them anymore. The eight-week course is followed by four refresher sessions every three months or so over the next twelve months.

At the heart of MBCT is learning to recognize the difference between what Teasdale calls the "doing" mode of thinking and behavior, which is purely automatic and habitual, and the more mindful "being" mode, which is conscious and involves metacognitive awareness—the ability to experience thoughts and feelings as transient phenomena.[24] Once participants have started to perceive these patterns of thinking more clearly, they can recognize when their mood is beginning to dip, but without adding to the problem by falling into old habits of rumination. "They learn to stand on the edge of the whirlpool and watch it go round, rather than disappearing into it," says Segal. "This helps break the old association

between negative mood and the negative thinking it would normally trigger." Like all mindfulness-based approaches, the program
teaches people to allow distressing emotions, thoughts, and sensations to come and go without trying to fight them, suppress them,
or run away from them. "They learn to stay in touch with the present moment, without being driven to ruminate about the past or
worry about the future," he says. This is the essence of mindfulness.

One of the techniques taught during an MBCT course that participants say they find particularly useful is the "three-step breathing space," which you can try for yourself by following the guide at
the end of this chapter. This is an informal practice that need not
necessarily be scheduled for a particular time of day. Rather, you
can use it to change the way you relate cognitively to stress or anxiety whenever they arise—on an overcrowded subway train, before
a meeting or an interview, or after a difficult encounter with a colleague. It has been called a "mini-meditation," because it needn't
last more than three minutes or so—and certainly not much longer
than the average office comfort break. The overall form of the
mini-meditation has been compared to an hourglass, because it
starts with a broad focus of attention, then narrows, then broadens
out again, with each step lasting around a minute.[25] Judith Soulsby,
who was Williams's research officer at the University of Wales on
the first clinical trial of MBCT, explains that the exercise is designed to break the feedback loop that normally perpetuates the
unpleasant physiological effects of stress. "We feel stressed and then
we stress ourselves even more by thinking how difficult everything
is going to be or how dreadful everything used to be," she says. "Just
taking that break and doing something that is soothing to the body
and calming to the mind is very helpful." In addition, she says, the
exercise contributes to the "decentering" that is at the very core of

mindfulness teaching. "The whole idea is to step outside our experience rather than being caught up within it—stepping outside and seeing, *aha!* that's a thought, that's a habit pattern, this is the emotion I'm experiencing at the moment and here it is in my body, it's in my chest or stomach or wherever. That gives us a different place to perceive what we're experiencing. We don't necessarily change what's happening, but we change how we relate to it."

Now in her seventies, Soulsby still leads courses and trains teachers at the university's Centre for Mindfulness Research and Practice (and regularly attends meditation retreats at Amaravati Buddhist Monastery). Results from the preliminary trial she helped to organize were published in 2000.[26] They were promising, and several more clinical trials followed over the next decade, adding evidence that MBCT is an effective way to prevent relapse into depression. In 2011, Jacob Piet and Esben Hougaard at the University of Aarhus in Denmark published a review that pooled the results of six randomized controlled trials involving nearly six hundred participants in total. They found that among people who had experienced at least one episode of major depression, taking an MBCT course reduced the risk by 34 percent that they would relapse, compared with those given a placebo or treatment as usual. For those who had gone through three or more previous episodes, the course reduced the risk of relapse by an impressive 43 percent.[27]

These trials and others had already persuaded the UK's health advisory body, the National Institute for Health and Care Excellence (NICE), to recommend first in 2004 and again in 2009 that MBCT be offered to people who are currently well but have experienced three or more previous episodes of depression.[28] Doctors usually treat an acute episode with an antidepressant, which the person will continue to take after they recover, for between six months and three years, to prevent a relapse. MBCT provides ongoing protection after this period is up, or sooner, if people decide they want to come off the antidepressant because of unwelcome side effects such as anxiety, agitation, or sexual dysfunction. "I think where MBCT is very effective is that it sequences seamlessly with antidepressant medication," says Segal. People get well by taking an antidepressant and then, after discussing the options with their doctor, they can join an MBCT group and receive a similar amount of protection even after discontinuing their medication. Another important advantage of mindfulness over drugs is that it teaches life skills that will continue to offer benefits, whereas an antidepressant only works for as long as a person takes the pills regularly. The benefits of MBCT go beyond relapse prevention, according to Willem Kuyken, a professor of clinical psychology at the University of Oxford and director of the Oxford Mindfulness Centre. "Mindfulness can make them feel more awake, more alert, more responsive in their relationships, more present to their partner and children."

Perhaps the most remarkable discovery to emerge from clinical trials of MBCT is that mindfulness offers the most benefit for adults who report having experienced severe adversity, abuse, or parental neglect during their childhood. In 2015, the most definitive trial of MBCT to date was published in *The Lancet*. Led by Kuyken, who was working at the University of Exeter during the trial, it involved a total of 424 adult patients with recurrent major depression who were followed for two years. Among patients who had reduced or discontinued their antidepressant medication and instead took an MBCT course, 44 percent relapsed into depression during this period, whereas 47 percent of those who carried on taking the pills had a relapse.[29] While this small apparent benefit of MBCT compared with standard drug therapy was not statistically significant, the advantage was clear-cut among patients who had experienced the most adverse childhoods. These turned out to be patients who had undergone more psychiatric treatment in the past, including more hospitalizations. The first onset of depression had been earlier, they had gone through more episodes, and had made more attempts to end their own lives. They were also more likely to have a family history of suicide and mental illness. Among this subgroup of participants, the relapse rate was 47 percent among those who came off their antidepressant and took an MBCT course, compared with 59 percent for those who continued to take the pills. Several other studies have seen similar results.[30] So, while not a cure-all, there is something about MBCT that provides measurable benefits for those who have suffered most from early childhood onward.

I asked Kuyken to explain what might be happening. "The premise of MBCT is that it is teaching people to become more aware of negative thoughts and feelings, cognitive reactivity and a tendency to ruminate. People with an abuse history are more likely to show these characteristics," he says. "Little triggers, both internal and external, can spark off all these thoughts of 'I'm no good,' 'People don't love me,' 'I'm going to be discovered as incompetent'—thoughts that can very quickly spiral into depression."

We don't yet know for sure what it is about MBCT that helps the people most vulnerable to depressive relapse become more resilient, protecting them from future episodes. Ongoing research by Kuyken and others into these mechanisms of risk and resilience could eventually suggest ways to fine-tune the therapy to make it more effective. The challenge will be to tease apart the benefits that most types of psychotherapy delivered through group sessions have in common—

such as making friends with fellow sufferers and reducing the perceived stigma of having a mental illness—from the benefits that are unique to MBCT. "For someone with a disorder like depression where a cardinal feature is feeling different, alone and defective in some way, just sitting in a room with fourteen people who look very much like you—a builder, a teacher, a doctor, somebody who lives on my street—describing the same experiences you're having is very powerful in terms of reducing stigma, in terms of enhancing a common sense of humanity," says Kuyken. But he says the data that he and his colleagues have analyzed so far suggest there are additional benefits that are specific to MBCT, and that these are mediated by increased mindfulness, changes in ruminative thinking, and enhanced levels of compassion, both for oneself and others.[31]

In 2014, shortly before a summit in London organized by *The Economist* entitled The Global Crisis of Depression, I interviewed one of the keynote speakers, Simon Wessely, president of the Royal College of Psychiatrists and a professor of psychological medicine at King's College London. The statistics outlined at the start of this chapter make grim reading. I asked him what makes so many of us vulnerable to depression, while others seem to sail through life untroubled by it. "Depression isn't a single thing and it doesn't have a single cause," he told me. "When I teach I say it almost every five seconds: repeat after me, 'Psychiatric disorders are multifactorial.'" These illnesses have complex genetic and environmental causes, he said, but went on to add that one of the proven ways to prevent depression in adults is to provide additional support to mothers who develop postnatal depression, a condition that increases the risk that their children will also suffer from depression in later life. "Your early environment and exposure to your mother—the relationship you have with your mother—influences child depression and adult depression. We know that." Globally, a fifth of all mothers develop depression either during pregnancy or the following year, making maternal depression the second leading cause of disease burden among women, the first being infections and parasitic diseases.[32] Several other environmental factors during childhood make people more prone to depression in later life, such as abuse, low socioeconomic status, and social isolation.[33, 34]

Brain-imaging research is providing tentative evidence that abnormal development of the default network during early childhood as a result of these types of trauma may be partly responsible for increased vulnerability to depression. We have seen that overactivity in a particular part of the network—the cortical midline structures, what I've been calling the brain's "Self app"—plays a crucial role in the excessive rumination that characterizes depression. Wandering minds tend to gravitate toward self-referential thoughts, and these are often tinged with negative emotions, so a failure to regulate the activity of the default network may mean these emotions increasingly intrude into everyday life. Imaging studies have revealed that the default network is still developing—its nodes are not fully connected up—in children seven to nine years of age,[35] so their environment could be playing a crucial role in this process. According to this hypothesis, having a particularly stressful, abusive childhood "scars" the network in some way during its development, with long-lasting consequences. Chronic stress during childhood could alter levels of nerve growth factors, for example, which in turn would influence connectivity in the network.[36] These changes might well make a person more prone to rumination and depression in later life.

This raises the intriguing possibility that mindfulness training retunes these faulty connections. We've seen that MBCT works best in those participants who report having had the most abusive childhoods. There is also evidence that for people who have been through several episodes of depression, taking an MBCT course helps them to ruminate less, and that this is what protects them from relapse.[37] More research will be needed, however, before we can say with any certainty what changes in the brain make some people more vulnerable to depression than others and what it is about mindfulness training that makes their illness less likely to recur. But a leading hypothesis is that by bringing activity in the default network under conscious control, meditation improves emotional regulation and mental well-being. It may turn out that the same mechanism, which I will explore further in chapter 10, "Wonderful and Marvelous," is the common factor underlying its efficacy in tackling depression, pain, stress, and anxiety.

Abnormalities in the default network have also been implicated in several other common mental illnesses, including bipolar disorder, psychosis, post-traumatic stress disorder, obsessive-compulsive disorder (OCD), and attention deficit hyperactivity disorder (ADHD), leading to hopes that meditation could prove beneficial in these conditions as well. It has even been suggested that meditation could help protect the brain from Alzheimer's disease, which is characterized by a progressive degeneration of the default network, possibly as a result of a lifetime's overactivity. I'll have more to say about this in chapter 11, "Mind Mirrors." At the time of writing, there is insufficient evidence (the kind that comes only from large, well-controlled clinical trials) that mindfulness is an effective preventive strategy or treatment for any of these conditions, though there has been some promising preliminary research.[38–41]

An early intervention that reduces teenagers' vulnerability to mental illness even slightly would have a massive impact on the future health of the population as a whole. Most mental health problems, including depression, develop during adolescence, with around 50 percent of adult mental health issues first appearing before the age of fourteen, and 75 percent appearing before the age of twenty-four.[42] Kuyken is now leading the biggest study ever undertaken into whether mindfulness training can protect teenagers from developing mental illness, which will last seven years and involve seventy-six schools in the UK. The study by scientists at the Oxford Mindfulness Centre and University College London has been funded by the Wellcome Trust to the tune of around £6.4 million. It will monitor the effects of the Mindfulness in Schools Project, a specially developed ten-week course that involves a thirty-minute lesson every week and up to twenty minutes' daily practice at home.[43] Starting in 2016, around 3,200 eleven- to fourteen-year-olds will take the course, while another 3,200 will receive standard personal, health, and social education lessons. Both groups will be monitored for their susceptibility to depression and their general mental well-being over the ensuing two years. A further 400 eleven- to sixteen-year-olds will be tested by neuroscientists at University College London before and after the course for their self-control and emotion regulation skills. Some of them will have their brains scanned to identify any associated changes.

"A mental training that teaches young people how to control their attention, recognizing when to respond to stimuli—particularly emotionally charged stimuli like exams or complex social dynamics—will really help them to navigate their way through school life, but also through life generally," says Kuyken. He was keen to emphasize that the program isn't just about negative challenges but also about helping young people to realize their potential in a sport or any other activity they enjoy. "For a teenager taking a penalty kick, it can be paralyzing to see their peers on the sidelines looking at them. Noticing that happening and being able to control their attention, focusing on what they're doing in that moment, is essentially a mindfulness skill."

Even in the absence of external distractions such as spectators at a sporting event, it feels as though the default mode network is constantly trying to lure our attention away from what we're meant to be doing. Unlike the brain's attention and sensory networks, which direct conscious awareness and process information coming from inside and outside the body to show us the world more or less as it is right now, the default network draws on memory to create mental simulations of the past, possible futures, the autobiographical self, and the perspectives of others. It conjures up holiday plans, conversations, and replays of last night's television, and endlessly repeats snatches of music. But its simulations don't always accord well with reality. Sometimes it offers up ruminations, anxieties, and delusions, which can tip the vulnerable into mental illness. As Milton wrote, "The mind is its own place, and in itself can make a heaven of hell, a hell of heaven."[44] On the evidence that it prevents depression in the most vulnerable people, perhaps a little more mindfulness could help us all adopt the more heavenly perspective.

A few hours after sunrise, Yasa's father showed up at the monks' encampment, having followed the distinctive imprints left in the mud by his son's golden slippers. He was almost as distressed as Yasa had been the night before. Nobody seemed to know where he was. Had Siddhārtha seen him? Before reuniting them, Siddhārtha told him the truth about suffering and the path leading to its cessation, then he explained that it would be impossible for Yasa to return to his old life because he was now free from all worldly attachments. All the wealth and pleasures of Varanasi no longer had any hold over him. Yasa's father was saddened to hear this, but he realized there was very little he could do or say that would change his son's mind. So he did what any right-thinking father would do in the same circumstances: he invited the Buddha to lunch.[45]

This was a pattern that would be repeated again and again over the ensuing weeks and months as word spread through the city about the inspirational teacher living under a banyan tree in the deer park with his growing band of monks. They came from the city to hear him preach the Dhamma, the radical new teachings that when fully realized were said to end the suffering that stalked everyone from birth to death—even the rich and privileged. Hundreds of women and men became lay disciples after hearing him teach, and many men (there wasn't yet an order of nuns) renounced their old lives, cut off their beards, shaved their heads, and donned yellow robes. They all took refuge in the Buddha, the Dhamma, and the Sangha.

According to the scriptures, there were now sixty-one enlightened beings in the world. At the end of the rainy season, Siddhārtha called them all together and addressed them. He said he was sending each one out individually—no two were to travel together—to spread the word far and wide, giving them authority to ordain novices and monks.[46] Go now, monks, he told them, "and travel for the welfare and happiness of the people, out of compassion for the world, for the benefit, welfare and happiness of gods and men."

Guided Meditation: Three-Step Reboot

This is also known as the Three-Minute Breathing Space, developed as part of mindfulness-based cognitive therapy (MBCT).[47] It's a handy way to dissipate stress and refocus your mind that you can take with you wherever you go. The meditation is short enough to fit into the average coffee or comfort break, making it ideal for easing the irritation, hostility, anxiety, and strain that can build up if you're having a challenging day.

When I first learned about this mini-meditation, I was reminded of the way an old office computer needs to be rebooted whenever its processor gets clogged by all the apps it is trying to run simultaneously. When you call IT, in all probability the first thing they say will be, "Have you tried turning it off and turning it back on again?" In spite of all the jokes, this is excellent advice, for both an old desktop computer and the human mind. Like a reboot, this meditation can clear away the clutter on your desktop, allowing you to see more clearly what's what. It will help you regain a little perspective and allow your brain to operate more comfortably, efficiently, and harmoniously.

The overall shape of the mini-meditation has been compared to an hourglass: its focus is initially broad, then single-pointed, then broad again. Each of the three steps can last around a minute, but allow yourself less or more time as necessary.

Step One—Becoming Aware

Whether you're sitting or standing, adopt a relaxed, upright, attentive posture. If possible, close your eyes. Check in to your mind and body, acknowledging whatever thoughts, feelings, and bodily sensations are present at this moment without attempting to change them in any way. Remind yourself that all these constituents of your present-moment experience are transient mental events.

Step Two—Gathering and Focusing Attention

Narrow your attention to focus on the sensation of the breath entering and leaving your body through your nostrils. Use the breath to anchor yourself securely in the present moment. Spend some time appreciating the sensation of your diaphragm and abdomen slowly rising and falling in synchrony. As you follow your breath entering and leaving the body, notice the tension in your muscles ease and the bones settle in their joints. Allow yourself to relish these sensations of relaxation. When the mind wanders, which it probably will, usher it politely back to the breath.

Step Three—Expanding Attention

Take your attention on a tour of the body as a whole: face, shoulders, chest, abdomen, legs, feet, hands, arms, shoulders. If there are any troublesome sensations of discomfort in a particular location, temporarily make these your focus. Remember that you are not trying to suppress unpleasant sensations but rather get to know and befriend them. What do they actually feel like, experienced simply as they are, without your making any attempt to explain or suppress them? Imagine you are breathing into the sensation and out again with each inhalation and exhalation. When you've made your peace with the sensation, widen your attention to the whole body and note how it feels in the here and now. This is where you want to be.

CHAPTER SEVEN

FIRE WORSHIPPERS

*Even a shower of gold cannot quench the passions: they
are wise who know that passions are passing and bring
pain in their wake.*

— The Dhammapada (translated by
Eknath Easwaran), verse 186

We are standing at the summit of a bleak, rocky hill on the west
bank of the Nerañjarā River in Uruvelā, just a few miles downstream from where Siddhārtha attained enlightenment. All about
us on the flat hilltop, occupying almost every square foot of bare
rock and red dirt, is a vast crowd of ascetics. There are at least a
thousand holy men, most of them wearing dhotis and with twisted
braids hanging from their left shoulders across their bare chests.
Some are clustered in groups absorbed in angry discussion, some
perch uncomfortably on boulders eating the remains of their

meager lunches, others pick their way carefully, barefoot, through the throng in search of a better vantage point.

These are fire-worshipping Jatilas, hermits who live in the forests beside the river practicing Yoga and the ancient Vedic rites. They tend the three sacred flames, burning sacrifices of brown rice to their gods at sunrise and sunset, and in the winter months they plunge into the biting cold waters of the Nerañjarā to cleanse their souls. Three colonies of the forest dwellers led by three charismatic brothers have come together for this meeting on the exposed hilltop. Most have the trademark matted, unkempt hair of the ascetic, but a few have freshly shaved scalps still flecked with dried blood. The pale skin stands out nakedly against the deep brown of their weathered faces. This shaving of scalps is a kind of betrayal, a repudiation of the ascetic way of life, but these men hold their peace and refuse to answer any questions.

Shaved heads are just one among many disturbing developments the fire worshippers have had to confront over the past few days. The leader of the largest colony, Kassapa, a highly respected brahmin, is rumored to have come under the influence of a sage from the Shakya clan in the north, a man who is not even brahmin. Most shocking of all, it has been whispered that Kassapa and his attendants have allowed the three sacred fires to go out and flung their ceremonial braids and the paraphernalia of fire worship into the river. Trusted witnesses saw the holy objects floating downstream. The Jatilas have assembled on Gayasisa Hill to demand an explanation.

A wave of silence breaks over the hilltop and everyone turns to discover its source. The three brothers, leaders of the three colonies, have just appeared at the top of the stone staircase that climbs the hillside from the plains below, and walking a few steps ahead of them is Siddhārtha Gautama. The Jatilas part before him as he makes his way toward the highest point. His palms are pressed together at his chest in greeting and he embraces them all with a smile. He looks much stronger than when Yasa met him several weeks earlier. There is more flesh on his bones and color in his cheeks, and he now wears a robe made from a single length of rough cloth dyed the color of the earth. He too has shaved his head. Helping hands hoist him onto a stone slab like a platform and a wide space is cleared. A thousand fire worshippers sit down as one and wait for him to speak.

"O Jatilas," he begins in a calm, authoritative voice, "everything is burning. The sense organs and the objects of the senses are aflame, perceptions are aflame, feelings are aflame, thoughts and consciousness are aflame. They are burning with the fires of craving, hatred and delusion, and as long as the fires find fuel on which to feed, they will continue to burn and there will be birth and death, decay, grief, lamentation, suffering, despair and sorrow. Brothers, a learned and noble disciple who realizes this truth will walk in the eightfold path of holiness. He will become wary of his eye, wary of all his senses, wary of his thoughts and ideas. When he no longer clings to these, the fires of craving, hatred and delusion will be quenched. He will be delivered from selfishness and attain the blessed state of nibbāna."[1]

For a long time, the fire worshippers sit silently contemplating what has been said. Then a few begin to murmur. Others join in and together they mouth words rhythmically in unison. As still more join in, it becomes clear that they're repeating the sermon, memorizing it. While they are absorbed in this task, Siddhārtha climbs down from the stone slab and makes his way toward the pathway down the mountainside. On either side, the heads of the Jatilas bow as he passes.

It could have gone either way. To tell a crowd of angry fire worshippers that "everything is burning" would prove to be either a psychological masterstroke or a diplomatic disaster. It turned out to be inspired because, according to the story, a thousand Jatilas became Buddhist monks after hearing what was to become known as the Fire Sermon—the third teaching or "discourse" delivered by Siddhārtha after his enlightenment. In the sermon, he is effectively comparing the five components of human existence—the physical world, our sense organs, perceptions, any pleasant or unpleasant feelings that arise, thoughts and consciousness—to bundles of firewood. He says they are all ablaze with delusion, hatred, and craving. According to Buddhist theory, these powerful forces are the cause of suffering. It starts with the delusion that each of us is an island or "Self" distinct from everything and everyone else in the world. This leads us to cling tenaciously to the things we want for ourselves and hate whatever is preventing us from getting them. The fire metaphor becomes clearer in the original written version of the story, in the ancient Indian language Pāli, because the word for the five components of human experience is *khandas*, which also means "heaps" or "bundles," and the word for "clinging" is *upādāna*, which is also "fuel." The literal meaning of *nibbāna* is to be "snuffed out" like a flame. So the lesson the Jatilas were meant to draw was that if they could stop clinging greedily to the objects of the senses and the mind, all their suffering would be ended. The "three fires" of craving, aversion, and delusion made a particularly cheeky analogy because of the three sacred, sacrificial fires that the Vedic priests kept permanently burning.

In a sense we are all fire worshippers, devoting our lives to feeding the flames of our passions. We were programmed by natural selection to pursue the pleasures of food and sex, to seek high social status, and to aggressively defend what we believe to be ours. In an environment such as a wealthy city—or indeed a royal palace—where food and other resources have become much more abundant than they ever were on the grassy plains of Africa where our species emerged, our passions burn out of control like a bush fire with the sheer abundance of fuel. In addition to providing plentiful food, the development of agriculture around 10,000 BCE would also eventually lead to the mass production of highly flammable materials to throw on the fire: addictive substances such as alcohol, nicotine, opiates, and cocaine. Being in thrall to an addiction is like riding a roller coaster of craving, gratification, and withdrawal, whether our passion is for something innocuous like caffeine; a substance that might easily kill us in the long run, such as tobacco; or a drug so addictive and behavior-warping—like heroin or crystal meth—that it will substantially reduce our life expectancy, either directly, through an overdose, or indirectly, as a result of violence or deadly viral infection from a needle. In our fool's paradise of abundant resources, we pay a heavy price even for our "milder" addictions. Smoking causes more than 5 million deaths worldwide every year,[2] and the economic toll of alcoholism is estimated at between 1 percent and 3 percent on average of every country's gross domestic product.[3]

Of course, it's not just the chemicals. Activities can be addictive too. I have a vivid childhood memory of the noisy games arcade on South Parade Pier at Southsea on England's south coast: the sweet smell of popcorn, the threadbare red carpet, the bright flashing lights and fairground sounds. It was there, on a family holiday in the 1970s, that my big sister and I got a taste for gambling. We were unable to play the one-armed bandits—being neither tall nor strong enough to pull down the handles—but we spent a thrilling hour placing bets on which wooden cutout horse would win in the derby game, feeding our precious holiday money into the slot of the machine. After each tantalizing, exciting race, all the horses would reverse at high speed back down the tracks to where they started, as if inviting us to try our luck again. Winning just once or twice in every ten games or so was all it took to hook us. The next day we came back for more and carried on playing the same machine until every last penny was spent. If our parents hadn't taken a stand and refused our urgent appeals for an advance on our pocket money, our family would have drowned in gambling debts by now.

I'm only half joking. This is a craving that starts young: a survey in 2013 found that 15 percent of eleven- to fifteen-year-olds in the UK had gambled in the past week.[4] And it's a habit that turns bad for a significant minority: among the adult population of the UK in 2010, just under 1 percent or around 451,000 adults were "problem gamblers," according to another survey.[5] Opportunities to gamble or play costly addictive games have proliferated in the past decade with the advent of fixed-odds betting terminals and of course the internet, smartphones, and tablet computers.

At first the drug or activity is immensely pleasurable, but with repetition it gets harder and harder to achieve the same "high" until—if we should sink so low—our only motivation is to relieve the gnawing discomforts of craving and withdrawal. The reward circuits of our brains give us the conviction that more will always be better—a bigger plate of tastier food, a more attractive partner, more money in the bank, a flashier car, more success in the eyes of the world—and yet we are never satisfied for long. We are on what psychologists call a "hedonic treadmill," when the novelty of any experience wears off and the "kick" we get from it subsides. The treadmill speeds up so we have to run faster, consuming more or having to find different sources of pleasure just to stay in the same place. It seems that the rewards we receive from anything pleasurable eventually revert to a preset norm (lottery winners and recently married couples will know exactly what I mean). This isn't a problem, but things can start going awry when we try to maintain the buzz of taking drugs, gambling, or perhaps even binge eating. We become "hungry ghosts," in the evocative imagery of Buddhist psychology, tragic creatures that roam one of the six lower realms of existence, unable to satisfy their constant craving. They are shown on the walls of temples with distended bellies but mouths so tiny and necks so slender that nothing ever reaches their stomachs. In the "hell garden" at Wat Phai Rong Wua in Suphanburi, Thailand, they are depicted in the form of carved wooden statues of emaciated white figures twenty feet tall with cadaverous grins.

In Buddhism, we are advised to let go of craving or desire, of the urge to hold on to pleasurable experiences and to be separated from painful or unpleasant ones. But the religion also recognizes some desires as wholesome. In 2013, in a dialogue about addiction between the Dalai Lama and scientists in Dharamsala, India, His Holiness observed that wholesome desires are indispensable for achieving any worthy objective, even the lofty spiritual goals of a monk. Without desire, he said, there would be no enthusiasm, no action, no progress. "Without desire I think then we would fall asleep!" he joked, slumping sideways in his chair to illustrate the point. Buddhists believe there is a middle way between total self-denial and overindulgence. After all, had the Buddha refused the bowl of rice milk that was offered to him in Uruvelā, he would have died of starvation and we'd all be none the wiser. The neuro-transmitter dopamine switched on the reward circuits of his brain, creating the motivation to eat and pursue his spiritual goal. Dopa-mine is released by the nucleus accumbens, a cluster of nerve cells beneath the cerebral cortex that neuroscientists have dubbed the "pleasure center." Addiction is what happens when a drug such as cocaine hacks into this system and keeps the switch in the "on" position, flooding the mind with a sense of intense well-being and the conviction that we can achieve almost anything. With repeti-tion, however, the intensity of the high wanes while the cravings strengthen and the mind's ability to make rational decisions weak-ens. The Dhammapada describes how compulsive urges grow until we're lost in "a forest of craving."

Up there on the hilltop the Buddha talked about the suffering caused by craving, but did he have a practical solution? Was the Fire Sermon no more than a nannying call for self-restraint or had the sage from Shakya discovered some kind of antidote? The clues could be in his brain: based on MRI studies of present-day monks, it's likely Siddhārtha's brain showed exceptionally low activity in his default mode network—not only when he was meditating but also in a state of rest. As we've seen, the default network fires up of its own accord whenever we're not focused on performing a par-ticular mental or physical task. Like an inbuilt time machine, it sends our minds ranging back into the past, rerunning particular events and social interactions, or ranging forward into the future to envision conversations and experiences that haven't yet happened. It also plays a central role in how we see ourselves and how we relate to other people.

The default mode can take us on extraordinary adventures of the mind: it is responsible for some of the greatest accomplishments of our species, making possible the feats of creativity and planning that are our hallmark. Without the brain's default mode, there would be no Pyramids, no Shakespeare sonnets, or Large Hadron Collider. But it has a darker side. In the previous chapter we explored its involvement in anxiety disorders and depression and its possible role in attention deficit hyperactivity disorder (ADHD), and we saw how mindfulness training helps to mute its activity. It turns out the default mode might also be implicated in drug addiction, though not through mind-wandering but rather because of the part it may play in the experience of being personally involved or "caught up" in a bodily sensation such as a drug craving. This is essentially what Buddhists mean when they talk about clinging or attachment, and they believe the principle applies equally well to thoughts and ideas. According to this view, the way to end suffering is to "let go" of our attachments. Of course, this insight isn't unique to Buddhism. The Hindu classics have a lot to say about the perils of attachment and the importance of renouncing them. Asked to sum up what he had learned in his life in fewer than twenty-five words, Mahatma Gandhi said: "I can do it in three!" Then he quoted the Isha Upanishad: "Renounce and enjoy."[6]

Compared with people who don't meditate, Siddhārtha's brain almost certainly showed much less activity in one of the principal components of the default mode, known as the posterior cingulate cortex. Other parts of his brain also would have exerted strong cognitive control over it. This is what Judson Brewer, a neuroscientist at Yale University School of Medicine and the University of Massachusetts Medical School, discovered when he compared brain scans of people who had been practicing meditation for more than ten years—clocking up an average of more than ten thousand hours—with those of matched controls who didn't meditate.[7] He found that in people who meditate, the activity of the posterior cingulate was more strongly correlated with activity in brain regions involved in cognitive control and detecting conflicts between different mental tasks (the dorsolateral prefrontal cortex and anterior cingulate cortex or ACC, respectively), suggesting that these regions are exerting a powerful influence over it. A study published in 2016 found that even in people who had never meditated before, just three days of intensive training was enough to strengthen cognitive control over the posterior cingulate cortex.[8] This in turn correlated with reductions in levels of a chemical in their blood called interleukin-6 that causes inflammation and is implicated in stress-related disease. Remarkably, the effects were still apparent four months after the training.

But what, if anything, do these changes in meditators' brains tell us about craving? We are still learning about the role of the posterior cingulate, but we do know that it is energy-hungry, well connected with other brain regions, and plays a part in learning and identifying with autobiographical memories. It is an integral part of the brain's cortical midline structures—the "Self app." If the posterior cingulate were a person, it would be a rather intense friend who insists on bringing out his family photo album at every opportunity. In addition to being intimately involved in creating our sense of self, we know that the posterior cingulate swings into action whenever our minds are wandering, and that it goes quiet when our attention is absorbed in a cognitively demanding task such as mental arithmetic or playing a computer game. But the posterior cingulate cortex has also been implicated in the complex brain mechanisms associated with craving. Researchers at the Medical College of Wisconsin in Milwaukee found the posterior cingulate was activated in cocaine users while they watched a film of two men talking about drugs and appearing to smoke crack cocaine, but not in control subjects who were not addicted to the drug.[9] Interestingly, a sexually explicit film activated the same area in both addicts and controls, leading the scientists to suggest it is part of the brain's normal craving response and not just involved in drug addiction. They point out that the posterior cingulate also fires up when people are thirsty.[10]

Brain surgeons in France have provided evidence of its involvement in nicotine addiction. In 2010 they reported the case of a thirty-five-year-old woman who had been smoking since she was seventeen. She was burning through more than two packs a day (her first smoke was usually within five minutes of waking) and had no intention of quitting. Then she had a stroke. She recovered well after emergency surgery and intensive care but reported a complete loss of interest in smoking, though she denied that having a stroke had motivated her to give up. A scan revealed a lesion in her right posterior cingulate cortex. The surgeons wrote that a year later she still wasn't smoking.[11] It was an isolated event—damage to this part of the brain is extremely rare—but, together with the other evidence, it puts the posterior cingulate cortex high on the list of suspects governing the complex mechanism of addiction. Brewer has speculated that, as a key node of the default mode network, it contributes to addiction by creating the sensation of being personally involved or "caught up" in a craving.[12] The Yale Therapeutic Neuroscience Clinic where he works uses mindfulness training to help smokers overcome their craving for nicotine, and researchers elsewhere are investigating its efficacy on other forms of drug and alcohol dependence.

We first encountered Professor Brewer in chapter 5, "The Man Who Disappeared," in the context of how the brain creates our sense of selfhood. He began practicing meditation in 1996, when he was going through a particularly rough patch after a relationship broke down. "I kind of came to it through my own suffering," he told me. "I was engaged to be married when I first started medical school and then we broke up. I couldn't sleep. I read this book about mindfulness by Jon Kabat-Zinn called *Full Catastrophe Living* and it was really, really helpful." I asked him what he means when he talks about the sensation of being "caught up" in difficult emotional experiences and the possible role of the posterior cingulate in both this and addiction. "When we're feeling nervous around someone, or when we're angry with someone or when we're scared, all of these involve getting caught up in experience," he explained. "Our field of view narrows and we're kind of 'clenched.' There are good studies suggesting that when you lie, when you're feeling guilty, when you're angry, when you're daydreaming, the posterior cingulate cortex is involved." He is not suggesting that the posterior cingulate is solely responsible—the more we learn about the brain, the more we realize that its operation depends on interconnected networks of different regions—but he says it could be a useful "sentinel marker" of this type of brain activity. It's worth noting that the *physical sensation* of craving—as opposed to the experience of being caught up in it—is associated with the brain's dopamine reward system centered on the nucleus accumbens.[13]

I asked Brewer how mindfulness could help with craving. He said the smokers who come to his clinic are taught to do something that goes completely against their instincts whenever they experience the urge to light up a cigarette. As with mindfulness-based stress reduction (MBSR), originally developed by Kabat-Zinn for patients with chronic, untreatable pain, they are told to focus all their attention on it, instead of avoiding or distracting themselves from the sensation.

"So smokers and people in pain just have to screw up their courage and face down the unpleasant sensations?" I suggest.

Brewer corrects me straight away. " 'Facing down' is typically how we do things in the West. You know, 'I'm gonna fight with you.' With mindfulness we actually teach people to become *curious*. If you can become curious about what your cravings feel like, suddenly they're not as unpleasant anymore, because curiosity itself is pleasant." This change in perspective—from taking things personally to viewing them dispassionately without getting involved or caught up in them—is fundamental to all mindfulness training. It is a shift in outlook that seems to be reflected in the brains of people who meditate. By using focused attention to temper the activity of the posterior cingulate cortex and the rest of the default mode network, they start to process bodily sensations in a more detached way. It is almost as if the "self" labels that our brains insist on sticking to every experience have lost some of their stickiness.

In Brewer's clinic they try to bring about this change among smokers who are desperate to quit but can't handle their cravings. "It's really about being able to notice these body sensations as they come up," Brewer explained. "Is it tension, is it tightness, is it burning? It's about realizing that in effect these body sensations are driving people towards getting cancer. Then they look at their cigarette packet and say, why do I do this?" Once smokers have learned that the cravings are simple bodily sensations rather than a personal imperative to feed their addiction, they will be better equipped to ride them out. At least that's the theory. Does it work? In a study published in 2011,[14] Brewer and his colleagues recruited eighty-eight nicotine-dependent men and women who were smoking around twenty cigarettes a day and wanted to quit (having tried and failed on average five times in the past). They randomly assigned them to two groups. One received mindfulness training and the other underwent the American Lung Association's Freedom From Smoking program, the gold standard treatment for smoking cessation, which involves teaching smokers strategies to help them control their behavior and reduce stress. Both types of therapy entailed eight training sessions spread over four weeks, plus home practice, with success measured by the number of cigarettes participants said they smoked daily before the treatment period, at the end, and thirteen weeks later. These personal reports were cross-checked by measuring the amount of carbon monoxide in their breath, which is a reliable, objective measure of how much they were actually smoking.

The results suggested that the smokers who received mindfulness training were significantly more successful at reducing their cigarette consumption and were more likely to have quit by the end of the treatment period. Three months later, 31 percent of the mindfulness group were still not smoking. This may not sound particularly impressive, but among those who had taken the Freedom From Smoking program, only 6 percent were still free from their habit three months later. People in the mindfulness group kept diaries reporting how often they practiced, and for every extra day of formal mindfulness practice per week, they smoked 1.2 fewer cigarettes daily on average. This is called a "dose response"—the more they practiced, the better the outcome—which in a clinical trial is a strong indicator that it is the treatment and not something else that is bringing about the observed benefits. For those on the Freedom From Smoking program, however, there was no correlation between the amount of home practice and the outcome.

Brewer and his colleagues were curious to know whether mindfulness was working the way they thought it was—helping smokers to handle their nicotine craving—so they conducted a further analysis of the data.[15] The smokers undergoing mindfulness training had reported their levels of craving and cigarette consumption before the study, at the end of treatment, and at follow-up interviews. When the scientists looked at the relationship between the number of cigarettes they were smoking and their levels of craving, a telling pattern emerged. At the start of the study there was a strong correlation between craving and smoking (in other words, the more they craved nicotine, the more they smoked). But by the end of the treatment period and at the follow-up points, the correlation was much weaker. What seemed to be happening was that rather than reducing their urge to smoke, mindfulness was helping the participants to tolerate their cravings without reaching for a cigarette. The level of craving itself started to wane only in the weeks and months after the treatment period ended, and then only among those who managed to remain abstinent. Among those who started smoking again, the cravings grew stronger.

That's the trouble with addiction: the habits have deep roots. Smokers who have quit are familiar with the urges that resurface whenever they are around other smokers; at particular times of the day, such as after meals or during coffee breaks, when they always used to light up; or when they come under stress. In the past, they have learned to associate these environmental and internal cues with the positive emotional effects of smoking—the social enjoyment, the calming sensations, the chemical buzz—and relief of the unpleasant sensations of craving. This is known as operant conditioning. In the 1920s, psychologist B. F. Skinner designed an experimental box for shaping the behavior of lab animals. He called it his operant conditioning chamber, but it later became known simply as the Skinner box. In essence, it is a microcosm of how environment determines behavior. First, Skinner showed he could train a rat to press a lever in order to receive a food pellet. Later, he trained it to do this whenever a light came on in the box. Pressing a lever in response to a light is not part of rats' "natural" behavior, just as setting fire to one end of a paper tube stuffed with dried leaves and sucking up the smoke isn't part of ours, but the animal had learned to associate this particular cue and that particular behavior with the pleasure of eating.

The intense pleasurable effects of an addictive drug or behavior are remembered in a part of the brain called the hippocampus, and the dopamine surge that they trigger primes the amygdala to associate them with particular stimuli. Psychologists call this a "conditioned response." For example, smokers learn from experience to associate all kinds of everyday internal and external cues— such as the physiological sensation of stress, the sight of a friend with whom they take smoking breaks, or even the taste of coffee— with the experience of lighting a cigarette and feeling their bodies suffused with stress-relieving, pleasure-giving nicotine. This may explain why although approximately 70 percent of smokers say they want to quit, fewer than 6 percent actually manage to achieve that goal each year.[16] A similarly poor success rate is seen among people trying to kick their addiction to illegal drugs or alcohol. In the US, only around 10 percent of people with these "substance use disorders" seek treatment, but even among these more motivated people, between 40 percent and 60 percent will have relapsed within a year.[17] Needless to say, public health services all over the world are keen to find more reliable ways to prevent relapse after treatment for drug addiction. According to Brewer, the problem with conventional strategies for treating addiction is that they teach people to avoid certain situations, or substitute one behavior with another (such as chewing nicotine gum or smoking an electronic cigarette), which leaves the underlying conditioning intact. This wouldn't necessarily be a problem, but the trigger situations may be so commonplace that they are impossible to avoid. So, while replacement strategies such as nicotine for smoking or methadone for heroin provide relief from craving, they don't prevent the sensations from returning when the old environmental cues inevitably present themselves in everyday life.[18]

By contrast, mindfulness therapy aims to take the sting out of craving so that the addict finds it progressively easier to resist. There is a close parallel here with exposure therapy for phobias. People who suffer from a phobia, such as an intense, irrational fear of spiders or social situations, typically cope by avoiding the object of their fear and trying not to think about it. With social phobia, for example, you might avoid parties; an arachnophobe might insist that his or her partner always deal with the spider in the bathtub. Both behaviors have the perverse effect of perpetuating the fear through "negative reinforcement"—you decide not to go to the party, and this immediately relieves your anxiety, making you more likely to avoid parties in future; your partner removes the spider, you feel better, and so next time you see a spider in the tub, you ask them to do it again. By contrast, treatment for phobias involves inviting the patient into a safe, controlled environment and gradually increasing their exposure to the thing they're most afraid of—social situations, spiders, or whatever—leading to what psychologists call "habituation" to the stimulus and the extinction of their fear response in the absence of any negative consequences. Similarly, mindfulness involves mentally turning toward unpleasant emotions and bodily sensations and observing them with dispassionate curiosity. This can lead to the astonishing discovery that the *unpleasantness* of the sensation diminishes when it is made the subject of detached awareness, and with time the sensation itself fades.

How would Siddhārtha have put it? He told the fire-worshipping Jatilas that "everything is burning" in the fires of craving, aversion, and delusion. Craving the object of our addiction can certainly feel like being burned alive, but by identifying personally with the physical sensations—by clinging to them—we only throw more fuel on the fire. It may feel as though the only escape is to reach for the bottle or cigarettes, and so the cycle of addiction starts again with the habits and their cues a little more deeply ingrained. In contrast, by making the cravings the focus of dispassionate awareness, we change our relationship to them and gain the insight that they will die down even if they are not satisfied. The fires of craving may still be smoldering, but given time—and in the absence of any more fuel—they will burn themselves out.

Brewer's research and other studies are providing preliminary evidence that mindfulness can help smokers quit, and there is good evidence that it could work for other addictions. In 2009, psychologists led by Sarah Bowen at the Addictive Behaviors Research Center at the University of Washington in Seattle started recruiting people who were being treated at a private drug rehabilitation clinic. They randomly assigned them to three groups. The first group received regular follow-up care based on the Alcoholics/Narcotics Anonymous 12-step program, including attending weekly meetings with patients and therapists for discussion and support. The second group were given a form of cognitive behavioral therapy (CBT), which included setting goals, social support, teaching them to assess high-risk situations for drinking or drug taking, and equipping them with cognitive and behavioral coping skills. The third group received a form of mindfulness training tailored to treat drug addiction, including some of the elements of CBT but also guided meditation and improving their awareness of physical, emotional, and cognitive phenomena. All three interventions were delivered at eight weekly group sessions.

The researchers continued to recruit people into their study until the middle of 2012, bringing the tally to 286, and published their results and conclusions in March 2014.[19] About 15 percent of the participants were abusing only alcohol, but around 80 percent were taking a range of drugs, including crack cocaine, heroin, and methamphetamine. It turned out that after six months of treatment, those who had received the standard follow-up care were significantly more likely to have returned to using drugs or drinking heavily compared with those who got CBT or mindfulness training to help them stay clean and sober. In the short term, CBT had the edge over mindfulness therapy, with these patients managing to stay abstinent for longer. But after a year, those trained to deploy mindfulness against their addictions reported significantly fewer days of substance use and significantly decreased heavy drinking compared with the CBT group. These findings were backed up by urine tests to confirm levels of drug and alcohol use. The psychologists concluded that mindfulness worked better in the long run because it improved people's ability to recognize and tolerate the discomfort associated with craving and negative emotions. Rather than simply avoiding high-risk situations where they might be tempted, they learned that craving is a straightforward bodily sensation that need not define them. They can experience it without being caught up in it.[20] In the absence of any further drug use, this appeared to weaken the ability of environmental cues to elicit craving.

When I phoned Bowen to find out more about the program, which they call mindfulness-based relapse prevention (MBRP), she explained that in addition to teaching clients how to live with their cravings without acting upon them, course leaders try to restore their enjoyment of ordinary activities. The tragedy of drug addiction is that experiences that are enjoyable for most people are no longer rewarding for someone with a history of drug abuse. Their dopamine reward system has been blunted by excessive release of the neurotransmitter as a result of taking drugs. "Things that were once naturally rewarding such as going on holiday and seeing the ocean are not rewarding anymore because the dopamine reward system is no longer triggered by them—it wants drugs. We ask clients, 'what are the things in your life that you miss?'" One of the aims of MBRP is to help people reconnect with activities they used to enjoy, restoring their sensitivity so they don't need to reach for something that will give them a huge "dopamine hit." The course teaches exercises that involve noticing and savoring experiences that are naturally reinforcing. "It might be something really simple, such as feeling your feet on the ground when you're walking, enjoying your cup of coffee in the morning and taking time to be with that experience, or hanging out with your kids and being present in a way that is going to be much more reinforcing and enjoyable." Bowen travels the world training people to provide MBRP. For example, she is working with psychologists at the Federal University of São Paulo in Brazil who provide the program to drug addicts in the city, and has trained MBRP teachers in the UK, Sweden, and Italy.

An increased tolerance for cravings as a result of mindfulness training may go hand in hand with improved self-control. As we have seen, brain scans of experienced meditators suggest that mindfulness meditation inhibits the activity of the default mode network, particularly the posterior cingulate cortex. It can also boost activity in prefrontal regions of the brain involved in decision making and emotional control, including the dorsolateral and orbitofrontal cortices and part of the anterior cingulate cortex (see Figure 4, page 165). In addicted individuals, these same regions are known to malfunction, impairing their powers of self-control.[21] A recent study found that two weeks of mindfulness training increased activity in the prefrontal and anterior cingulate cortices of students, which suggests that meditation may help counter this damage.[22] It also reduced the amount they were smoking by an impressive 60 percent. In a control group of smokers who underwent relaxation training, there was no reduction in cigarette consumption by the end of the study and they had no changes in these brain areas. There is a remarkable twist to this tale, though, because the students were not aware that the study had anything to do with smoking. When they were recruited, they were told it was all about learning to cope with stress. And yet their cigarette consumption fell by more than half—a success rate that would be the envy of anyone running a smoking cessation program. As any smoker who has tried to quit will tell you, consciously trying to kick the habit can have the perverse effect of making the cravings even worse in the short term. Thought suppression is notoriously counterproductive—try really hard not to think about lighting up a cigarette and soon you won't be able to think about anything else. If these results are confirmed in future studies, it would suggest that taking up meditation and not obsessing quite so much about quitting might be a better strategy. There has also been some research suggesting that mindfulness meditation works better than thought suppression for people who want to reduce their alcohol intake.[23]

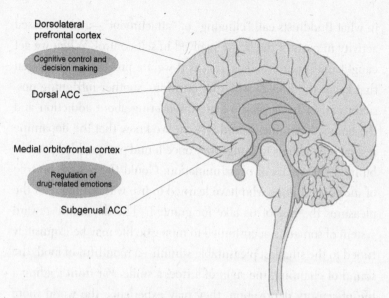

Figure 4. Self-control. In people addicted to drugs, brain activity is abnormal in prefrontal regions associated with self-control, impairing their ability to resist cravings. Imaging studies have found reduced activity in the dorsal anterior cingulate cortex (ACC) and dorsolateral prefrontal cortex, which are involved in top-down cognitive control over behavior, and in the medial orbitofrontal cortex and subgenual ACC, which handle the subconscious regulation of emotions. By boosting activity in these regions, mindfulness training may help addicts reduce their dependence on drugs.

Our peeks into "Siddhārtha's brain" hint at some of these beneficial spin-offs of meditation: enhanced willpower and a reduced tendency to get emotionally "caught up" in unpleasant sensations such as craving. In his sermon to the fire worshippers on Gayasisa Hill, he presented an insightful metaphor of the human condition: "Everything is burning . . ." MRI scans of his brain, like those of similarly experienced meditators, would no doubt have revealed distinctive structural and functional changes, including decreased activity in the posterior cingulate cortex—which may play a role in what Buddhists call "clinging" or "attachment"—and increased activity in parts of the brain involved in self-control. When we get caught up in our cravings, it is as if we are providing the oxygen that keeps the fires of suffering burning. Another intriguing possibility arises from what we are discovering about addiction and the body's dopamine reward system. We know that the dopamine response of addicts to things they once found pleasurable becomes blunted with years of overstimulation. Could the reverse be true of monks and nuns who have learned to live without many of the pleasures the rest of us take for granted? The dopamine reward system of someone accustomed to monastic life may be exquisitely tuned to the smallest pleasurable stimuli—a mouthful of food, the sound of chanting, the sight of a tree, a smile. Far from leading a life of sensory deprivation, they may experience the world more intensely, more vividly than we can imagine.

CHAPTER EIGHT

A DRUNK ELEPHANT

Speak quietly to everyone and they too will be gentle in their speech. Harsh words hurt and come back to the speaker. If your mind is still, like a broken gong, you have entered nirvana, leaving all quarrels behind you.

—The Dhammapada (translated by
Eknath Easwaran), verses 133–34

"They're coming!" The long, dusty street is deserted but the people of Rājagaha, ancient capital city of the kingdom of Magadha, crowd their rooftops and balconies and poke their heads from windows to catch a glimpse of the living stream of ocher, bright orange, and yellow flowing down the road from the Bamboo Grove. Like a barefoot army, thousands of monks from eighteen local monasteries are issuing in single file from the gates of the grove in the shadow of Vulture Peak. Leading them at an unhurried pace is an elderly monk cradling an alms bowl.

From the comparative security of their houses, the townsfolk are just beginning to enjoy the festival atmosphere when the air is split by a terrifying sound. At the far end of the street, partially hidden by great clouds of dust, a monstrous bull elephant is trumpeting furiously, trunk raised and ears forward as he charges from side to side, overturning and trampling carts to splinters under his feet, barging through the flimsy shop fronts like undergrowth in the forest. Each of his tusks is as long as a chariot pole and his head as solid and unforgiving as a battering ram. This savage beast is the manslayer Nālāgiri, one of the king's fighting elephants. Ever since he was dragged from his mother's side while still a calf, Nālāgiri has borne a murderous grudge against these arrogant creatures that swagger on two legs and stab with sharpened sticks. He has killed many times, and not only on the battlefield. His keepers usually hobble him with heavy iron chains about his legs, and every morning they bring him a bucket of distilled palm wine, or arrack, which befuddles and pacifies him. But today a monk named Devadatta arrived at the stalls claiming to be a relative of the king. Issuing threats of demotion and promises of advancement, he ordered Nālāgiri's keepers to give the murderous creature two buckets of the fiery liquid to drink instead of one; then they loosed his chains and used their goads to drive him out into the street.

Devadatta, in truth, is unrelated to the king but is in fact Siddhārtha's cousin and bitter childhood rival. Years earlier, after a visit by the Buddha to his hometown, this man and Ananda—who has been the Buddha's faithful attendant for decades now—were ordained into the rapidly expanding Sangha. Devadatta has long harbored an ambition to wrest the Sangha from the Buddha's control. A brilliant orator, he recently used the occasion of a Dhamma talk attended by the king and a large gathering of monks to call on the seventy-two-year-old to retire—and was very publicly rebuffed. Humiliated and driven by ambition, he dispatched archers to kill his rival, but the soldiers could not bring themselves to harm the Buddha, who then converted them all. When Devadatta took matters into his own hands and hurled down a boulder from Vulture Peak as the Buddha walked below, it only grazed his foot. Finally he had found an assassin who would not be swayed by the wise words and charisma of his victim.

On the main street of Rājagaha, the procession has come to a halt as senior monks make a final desperate effort to persuade the Blessed One to get out of the path of the rampaging elephant, but he will not be moved.

"Then let the elephant kill me first!" cries Ananda, stepping in front of his master.

Siddhārtha brushes him aside: "Don't be silly, Ananda."

The sounds of destruction have become alarmingly loud, but great clouds of dust hang over everything, obscuring the monster. Everyone is taken by surprise when Nālāgiri emerges from the swirling clouds like a mountain just twenty yards from where the Buddha is standing impassively in the middle of the road. Behind him the monks have lined up in dense ranks and sat down as one. Now they begin to chant. The beast pauses, ears twitching, but all he can see is a vague field of color. Then the elephant notices movement out of the corner of one of his wildly staring eyes: a woman running for her life. He raises his tail and charges. A child falls screaming from her arms as she flees and, hearing the piercing sound, the beast breaks its stride and shudders to a halt, towering over the infant lying helpless on the ground.

The Buddha chooses this moment to call softly, "Nālāgiri!"

Turning his great head and shaking his ragged ears as if he

has only just noticed the old man in the middle of the road—so still and quiet—the elephant turns away from the child and approaches the Buddha. Suddenly all the fight goes out of him and he falls to his knees as if it were the king himself standing there. Siddhārtha puts out his hand to stroke the broad forehead and says a few words about kamma and the suffering we will reap in the next life as a result of our wrongdoing in this one. When he has finished, Nālāgiri sucks up some dust from the Blessed One's feet with his trunk and sprinkles it over his own head. Then with a mighty heave he rises to his feet and returns placidly to his stall.[1-3]

For millennia, trained elephants have been revered in India not only for their strength but also for their intelligence and loyalty. In Siddhārtha's lifetime, the animals were prized fighting machines that could determine the course of a battle and the fate of kingdoms. They were also highly valued as patient beasts of burden for lifting and dragging immense weights such as tree trunks, and were still being used for this purpose in the Indian timber industry for most of the twentieth century. Nonetheless, elephants have always evoked fear and distrust because of their tendency to run amok when provoked. The World Wide Fund for Nature estimates that despite declining numbers, wild elephants kill up to two hundred people a year in modern India.[4] In the Buddha's day, this dual nature—the patient endurance of a well-trained animal and the latent violence of a wild or poorly trained one—made the elephant a perfect metaphor for the mind, symbolizing the best and worst in human nature. In Buddhist legend Mara, the demon of death, rides an elephant "150 leagues high," but the Buddha himself was often compared to a great elephant: he had tamed the powerful but unruly mind. In the story retold above about Nālāgiri, from

The Jataka Tales, his encounter with Nālāgiri is framed almost as a prize fight, an epic confrontation between "the lord elephant Buddha and this elephant of the brute world."

You don't have to look very far to find elephant analogies in Buddhist scripture. The Dhammapada contains several passages that compare the mind to an elephant that must be trained to realize its full potential:

> *Patiently I shall bear harsh words as the elephant bears arrows on the battlefield. People are often inconsiderate. Only a trained elephant goes to the battlefield; only a trained elephant carries the king. Best among men are those who have trained the mind to endure harsh words patiently. . . .*[5]
>
> *Long ago my mind used to wander as it liked and do what it wanted. Now I can rule my mind as the mahout controls the elephant with his hooked staff. Be vigilant; guard your mind against negative thoughts. Pull yourself out of bad ways as an elephant raises itself out of the mud.*[6]

Attaining mastery over unruly emotions is a central objective of Buddhist meditation. The areas of Siddhārtha's brain associated with emotional regulation would have been transformed by decades of practice. Brain scans of similarly dedicated modern practitioners suggest that in the early years, there would be an increase in the thickness of parts of his cortex involved in attention and awareness of internal bodily states, vital for emotional control, and these changes would be maintained for much of his life.[7] "Gray matter"—essentially nerve cell bodies and their profuse connections—would become more tightly packed in several regions, including his insula and somatosensory cortex, which monitor internal and external stimuli respectively, and his hippocampus, which plays a vital role in memory and the unlearning of fear conditioning.[8, 9] As a child and later as a young man, there may have been heightened electrical activity in the frontal part of his right hemisphere compared with the same area on the left, suggesting a somewhat negative outlook on life,[10] but after intensive meditation in his thirties this pattern reversed. Alpha waves now would pulse with greater intensity on the left side of his brain than on the right.[11] Thanks to the untiring work of the DNA repair enzyme telomerase, his chromosomes would have also been in excellent shape for a man his age. This would suggest not only that he gets plenty of exercise and has a good diet, but also that he is a stranger to chronic stress.[12-16] In fact his heart rate, skin conductance, and the levels of a hormone called cortisol in his saliva—physiological measures of stress—likely returned to normal just minutes after his potentially fatal encounter with the murderous Nālāgiri.

This apparent ease of emotional regulation would be explained by changes in many parts of his brain over the previous four decades. In the early years of his meditation practice, both gray matter density and activity likely increased in specific areas of his anterior cingulate and prefrontal cortex, suggesting deliberate, top-down regulation of powerful emotions such as anger, excitement, anxiety, and fear. But in later years this activity would have peaked and then started to decline, reflecting increased bottom-up, unconscious regulation.[17] In other words, his powers of emotional control became automatic, almost effortless. "Letting go" was second nature to him. If you were to plot on a graph the activity changes in these cortical regions in the first ten years or so since he began to meditate, you would get an upside-down, U-shaped curve. Neuroscientists recognize this as an effect of dedicated training that occurs with growing expertise in a wide range of human activities, from athletics to mathematics, from juggling to reasoning, from knitting to carpentry. In the early days, a student or apprentice's work requires conscious effort and concentration, but with practice and as changes in the central and peripheral nervous systems are optimized and become hardwired, there is an increasing ease of performance and a decreasing need for conscious control. The necessary thought processes and behavior have become habitual. This is what neuroscientists mean by "muscle memory." So, for example, when a high jumper first learns how to perform the Fosbury Flop—the peculiar, arch-backed, headfirst technique for clearing the bar—they must actively monitor and control what their body is doing at every stage of the run-up, takeoff, and flight. But after hundreds or thousands of repetitions, the whole sequence of movements becomes almost completely automatic, allowing for more subtle competitive factors to come into play. In the same way, a pianist learning Prokofiev's Third Piano Concerto will start by sight-reading this fiendishly fast piece, playing slowly and with little dynamic expression, but after many hours' practice will perform it with increasing ease. It is as if the hands themselves were learning the piece, allowing the pianist to focus more and more on the nuances of feeling that can be expressed by the music.

The same process may occur as people learn, through meditation, how to control troublesome emotions. When psychologists talk about emotional regulation, they mean the strategies deployed to influence which emotions arise and when, how long they last, and how they are experienced and expressed.[18] How, for example, does someone terrified of public speaking learn to master that fear?

What inner resources allowed Siddhārtha to stand his ground against a rampaging elephant? The four principal emotion regulation strategies recognized by psychologists are avoidance, distraction, suppression, and reappraisal. Avoidance involves steering clear of situations that provoke strong sensations such as fear or craving. As we saw in the previous chapter, this leaves the underlying conditioning intact and may even strengthen it. Avoidance is undoubtedly the most sensible strategy where an angry elephant is concerned, but not for something such as an intense fear of job interviews, which, if left unaddressed, might seriously impede one's career. Alternatively, distracting yourself by imagining the interviewer stark naked—a popular strategy—might at least stop you getting up from your chair and fleeing from the room. Suppressing unwanted emotions is perhaps the least successful strategy of all because it is liable to backfire.[19] According to the "ironic process theory" proposed by Harvard psychologist Daniel Wegner in the 1980s, active suppression of an unpleasant thought or emotion makes it more likely to pop up at just the wrong moment, when your brain is overloaded by a stressful situation or a demanding cognitive task.[20] Fyodor Dostoevsky formulated the classic example of this phenomenon some 150 years ago. "Try to pose for yourself this task: not to think of a polar bear, and you will see that the cursed thing will come to mind every minute,"[21] which is why ironic process theory is known more informally as "the white bear problem."

An explicit aim of the mindfulness-based stress reduction (MBSR) program developed by Jon Kabat-Zinn at the Massachusetts Medical Center, which I introduced in chapter 4 in the context of pain relief, is to help people develop skills for managing runaway stress and anxiety. In eight weekly group sessions lasting up to two hours, plus a "day of mindfulness" in the sixth week, participants learn sitting and walking meditation, yoga exercises, and the body scan (see "Guided Meditation: Body Scan," page 236). With the help of an audio recording, they also practice meditation at home (and are encouraged to perform everyday activities, such as eating, washing dishes, and taking a shower, more mindfully). The idea is that focusing attention on the breathing relaxes the body and soothes the mind. This is the "calm" stage of meditation. Thoughts, emotions, and bodily sensations are then held up to the mind's eye as they arise and investigated dispassionately, without any reaction to them. This is the "insight" stage. Reappraisal plays a large role in mindfulness, not only where internal distractions are concerned, but also by helping people to reconstrue intrusive external stimuli as more benign than their habitual reactions had been leading them to believe. Our tendency is to take every little distraction personally. Several years ago, when I made my first faltering attempts to meditate at home, distraction was a major problem. Early every morning after breakfast, enthroned in my favorite armchair, my eyes tightly closed, I'd find the electrical whine of a hair dryer starting up in a neighboring flat or the creak of floorboards overhead as the folks upstairs got out of bed intensely annoying. How inconsiderate of them! Suddenly my mind was all over the place. Recognizing these distractions for what they really were—simply other people like me starting their day—I eventually managed to loosen their emotional grip on me.

At Amaravati Buddhist Monastery in the UK, they still recount a story about the Thai Forest monk Ajahn Chah's first visit to the UK in 1977, accompanied by Ajahn Sumedho and another Western monk trained by the meditation master in Thailand. They had come on the invitation of the Hampstead Buddhist Vihara in north London. This was an important occasion, the visit of a revered teacher from the Far East to one of the first struggling outposts of Buddhism in the West. It was a hot, sticky summer's evening, and Ajahn Chah was leading them all in meditation in a cramped, stuffy room. Everyone was doing their level best to concentrate, but the temperature was rising and the air was throbbing with the sound of rock music from a pub across the street. Someone opened a window to let in some cooler air, only to close it again minutes later when the noise became insufferable. A delegation was even sent across the road to ask the pub's staff to turn the music down, but to no avail. All the while, as everyone else fretted and sweated, Ajahn Chah sat calmly meditating, only ringing the bell after they had endured this sensory onslaught for an hour and a half. Immediately several people began to apologize for the noise. Ajahn Chah finally spoke and an interpreter translated: "You think the sound is annoying you, but actually it is you that is annoying the sound. The sound is just what it is, just the air vibrating. It is up to us whether we start an argument with it."

As the Dhammapada puts it: "If your mind is still, like a broken gong, you have entered nirvana, leaving all quarrels behind you." Of course, to leave all quarrels behind is easier said than done. Starting arguments comes all too naturally to our reactive human mind—we wallop that gong at the slightest provocation. How could meditation help? The idea is not to numb the mind so it becomes impervious to emotion. Rather, the principle is that present-moment awareness increases our sensitivity to the bodily sensations associated with potent emotions such as anger or irritation, signaling the need for cognitive control. Then, with reappraisal and acceptance, the sensations ebb away quite naturally.

In the metaphorical language of the Fire Sermon, the flames die down in the absence of fuel to feed them.

By presenting volunteers with evocative pictures, sounds, or words as they lie in a scanner, neuroscientists investigate how programs such as MBSR influence how the brain processes emotion. For example, participants might view pictures of faces with aggressive, neutral, or friendly expressions, or hear sounds such as a woman screaming, the hubbub of a busy restaurant, or a baby cooing. They are instructed to address these stimuli either with mindfulness or with their ordinary state of mind. The research suggests that mindfulness increases activation in the insula and somatosensory cortex in response to emotional provocations.[22] These regions are involved in monitoring sensory information coming from our internal and external environment respectively. As we have seen in previous chapters, directing attention toward unpleasant bodily sensations also plays a major part in the efficacy of mindfulness for pain relief and treating drug addiction. An increased focus on the body's responses to particular stimuli may lead to greater awareness of one's emotional life, which is a prerequisite for gaining any kind of control. Troublesome emotions such as fear, anger, or craving can then be accepted without reacting to them, or reappraised as less threatening. Rather than being swept along by emotions, one observes them dispassionately, recognizing them as transient mental events that do not necessarily tally with reality.

Mindfulness is also known to boost activity in a part of the prefrontal cortex associated with voluntary emotional control called the dorsolateral prefrontal cortex (see Figure 5, page 181).[23] This region has extensive connections with other parts of the brain, not only muting activity in the default network (which as we have seen is involved in the experience of getting "caught up" in emotional states) but also in the amygdalae, structures buried deep inside the brain, one in each hemisphere. The amygdalae are a key component of the limbic system, the evolutionarily ancient brain network responsible for emotion and memory. Each one works like a smoke detector, setting off an alarm whenever it picks up signs of danger in the environment. Unlike a smoke detector, however, it can differentiate between stimuli that have caused us trouble in the past and those that have turned out to be harmless. Its job is to distinguish threat from safety by turning our attention toward emotionally significant stimuli in our surroundings and probing the significance of ambiguous ones.[24] To do this it stores memories of pleasant and unpleasant events that have followed particular stimuli. But it seems our long-term well-being depends critically on how quickly activation in the amygdala returns to normal after we encounter an emotional stimulus.[25] A wide range of disorders of emotion regulation, including depression, post-traumatic stress disorder (PTSD), social phobia, obsessive-compulsive disorder (OCD), and anxiety involve reduced prefrontal activation and chronic overactivity of the amygdala. To extend the smoke detector analogy, it's as if something has triggered the alarm, which then carries on ringing long after the smoke has cleared.

There is evidence that one of the factors underlying the effectiveness of mindfulness for treating anxiety and depression is that it increases activity in parts of the prefrontal cortex involved in emotional control, which in turn "downregulate," or damp down, amygdala activity. However, much work remains to be done to clarify exactly how mindfulness helps people to improve how well they handle emotions.[26] One particularly surprising finding has been that highly experienced meditators register *less* prefrontal activation in response to emotional stimuli. This may be because they have developed an accepting stance toward their emotions so that they no longer need to exert conscious, top-down cognitive control over them. So, while meditation novices use cognitive reappraisal to deal with strong emotions, adepts with thousands of hours of experience may have developed an automated response of *non*-appraisal—in other words, acceptance.[27] So when the smoke alarm goes off, while the rest of us are standing on a chair flapping a tea towel to make the wretched thing stop (downregulation), experienced meditators—like Ajahn Chah meditating serenely while the air around him throbbed with rock music—are content to let it ring (acceptance). They know it will stop once the smoke has cleared.

As we have seen, the objective of mindfulness is to turn one's attention toward troublesome sensations and feelings in a spirit of open-minded curiosity, rather than ignore them, try to squash them, or get caught up in them. Buddhists believe this mental attitude facilitates the "fading" or "cessation" of emotions such as craving, anxiety, hatred, and anger. In the language of modern psychology, repeated exposure to a fear-provoking stimulus without any negative consequences leads to the "extinction" of the fear response. Given a chance, the brain is capable not only of learning to fear particular situations but can also *unlearn* that response. If it didn't have this capacity to unlearn old fears, we would never pluck up the courage to leave the safety of our homes in the morning. As we saw in the previous chapter, exposure therapy for phobias such as arachnophobia harnesses this natural mechanism: the patient is exposed in small, easy steps to the thing that frightens them (spiders), gradually building up a feeling of safety in its presence. Mindfulness meditation can be viewed as a kind of internalized exposure therapy. There is growing evidence from neuroscience that the same brain mechanism comes into play in both mindfulness and exposure: the same network of regions involved in the extinction of conditioned fear—the hippocampus, amygdala, and part of the anterior cingulate cortex—is also influenced by mindfulness meditation. The hippocampus plays a part in both memory and emotion regulation, and, as previously mentioned, the amygdala has a central role in establishing fear responses. It seems the anterior cingulate cortex and hippocampus are vital for extinguishing them and keeping it that way.[28] Research suggests that meditation boosts connectivity between prefrontal regions and the amygdala, increases the density of gray matter in the hippocampus and bulks up the anterior cingulate cortex (see Figure 5, page 181).[29] We also know that exposure to chronically high levels of stress hormones *shrinks* the hippocampus, which is a characteristic of both major depression and PTSD.[30] By keeping a lid on anxiety as we go about our lives, with all their inevitable false alarms and minor mishaps, a mindful attitude may shield the hippocampus from stress-related damage and thus protect against mental illness.

In the past, the ultimate test of whether meditation works as advertised to improve emotional stability was considered to be whether improvements correlated with increases in "trait mindfulness." Psychologists researching mindfulness have tried to capture this elusive trait by handing out questionnaires for subjects to fill in before and after taking an MBSR course. Participants check a box on a scale from "almost always" to "almost never" in response to statements such as "I could be experiencing some emotion and not be conscious of it until some time later"; "I tend not to notice feelings of physical tension or discomfort until they really grab my attention"; or "I rush through activities without being really attentive to them." However, these questionnaires have fallen into disrepute in recent years after yielding some very peculiar results.

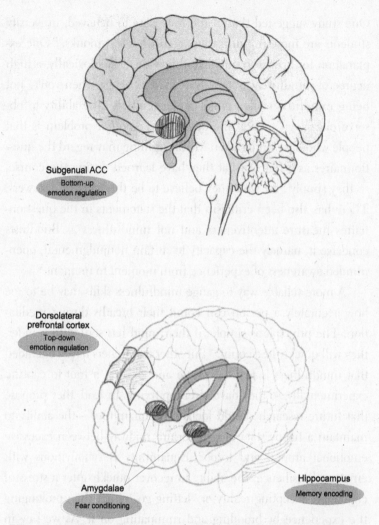

Figure 5. Emotion regulation. A network of brain regions responsible for un-learning fear responses, including the hippocampus and part of the anterior cingulate cortex (ACC), is boosted by mindfulness meditation. The same regions are activated during exposure therapy for reducing the irrational fears that are the hallmark of phobias. Mindfulness also increases activity in the dorsolateral prefrontal cortex, which plays an important role in volitional, top-down emotion regulation by inhibiting activity in the default mode network and amygdalae.

One study suggested that, if the tests were to believed, university students are more mindful than Thai Buddhist monks.[31] One explanation for bizarre results like this is that, paradoxically, a high degree of mindful awareness is needed to notice when you're not being mindful—it takes good "metacognition," the ability to observe one's own mental life objectively. Another problem is that people who have undertaken MBSR training may regard the questionnaires as a test of what they have learned during the course, so they simply give what they believe to be the "correct" answers. There has also been criticism that the statements in the questionnaires measure attentiveness and not mindfulness as Buddhists conceive it, namely the capacity to sustain nonjudgmental, open-minded awareness of experience from moment to moment.[32]

A more reliable way to gauge mindfulness skills may be to see how accurately a person can count their breaths during meditation. The principle is simple: if their mind has started to wander, they will quickly lose count.[33] But other researchers have concluded that mindfulness is just too subtle and slippery a trait to capture experimentally, so they have given up trying. Instead, they propose that future research should focus on "equanimity"—the ability to maintain a happy state of equilibrium midway between excessive emotional arousal and torpor. Equanimity is synonymous with emotional resilience: the ability to recover quickly after a stressful challenge, accepting reality or "letting go" rather than prolonging the experience by brooding and ruminating on it. As we saw in chapter 6, "Golden Slippers," mindfulness-based cognitive therapy (MBCT) was designed to foster this ability in people who have experienced several episodes of major depression, in order to prevent another relapse. Equanimity and emotional resilience are explicit goals of meditation, but unlike mindfulness, they can be measured physiologically by recording how quickly factors such as levels of stress hormones, heartbeat, and breathing return to normal after a frightening experience.[34] Rather than letting loose a drunk elephant in the lab, scientists might give their subjects stressful tasks to perform, such as delivering a speech in front of a video camera and two stony-faced scientists holding clipboards, then perform a tricky piece of mental arithmetic under their gaze. Physiological measurements are taken before the stress challenge and at intervals afterward.

Some of the most intriguing mindfulness research has involved more long-term physiological measures of stress and provides an insight into how meditation might benefit physical health as well as mental well-being. In a classic study led by Jon Kabat-Zinn and Richard Davidson, employees at a biotech company in Madison, Wisconsin, were randomly assigned to meditation and nonmeditation groups. Twenty-five took the eight-week MBSR course while sixteen were asked to wait for their chance to take it. After the eight weeks were up, to test the responsiveness of their immune systems, they were all given a flu vaccination. The amount of antiviral antibody in their bloodstreams was then measured at two intervals around five and nine weeks later. On three occasions during the course of the study, the electrical activity of their brains was also recorded using electrodes glued to their scalps (known as electroencephalography, or EEG).[35] It was found that the people who took the mindfulness course had significantly greater electrical activity in their left prefrontal cortex than the controls at the end of the training period and four months later, whereas there had been no difference between the two groups at the start of the study. Previous research has suggested that people with relatively greater prefrontal activation on the left side of their brain have sunnier dispositions and recover more quickly after stressful events.[36] In other words, they are more emotionally resilient. The MBSR group also reported feeling less anxiety by the end of the course compared with the controls, their overall mood appeared to have improved, and this effect was sustained even four months later. Most surprisingly, perhaps, the people who had learned how to meditate had a stronger response to the flu vaccine: the amount of antiviral antibody in their bloodstream was significantly greater than in the controls four and eight weeks after they were vaccinated. What's more, the meditators who had the strongest immune response to the vaccine also had the greatest increase in electrical activity in their left prefrontal cortex.

More research will be needed to confirm these results, but if they are to be believed, mindfulness meditation not only increases emotional resilience and mental well-being but also strengthens the immune system. By making people more emotionally resilient, mindfulness meditation may shield the immune system from the effects of chronic stress, which is known to suppress immune responses to pathogens such as viruses.[37] One common misunderstanding about meditation is that it numbs the mind to all emotion, both positive and negative, like a powerful tranquilizer. In reality, the objective is not to dull the emotions but to make people more alive to them at a visceral level, which, as we have seen, is a prerequisite for effectively regulating them, whether that is done deliberately or automatically. It would be a mistake to think that Siddhārtha felt no fear at the sight of an enraged elephant bearing down on him. A truly fearless person would be unlikely to survive very long, either in his world or ours. Even if they managed to avoid getting trampled or eaten, sooner or later they would be run over by a truck.

• • •

Molly shoulders open the heavy door, pokes her crutches through the gap, and awkwardly levers her body into the waiting room, simultaneously trying to maintain her balance while keeping her injured foot in its clumsy surgical boot off the floor. She winces as she makes her way toward the seating area in one corner of the room, though there isn't a single free seat there for her. She comes to a halt just a few feet from where three men sit, leans against the wall and sighs— evidently in pain. The men are around twenty years of age. One carries on reading his magazine as if he hasn't seen her. Another is fiddling with his mobile phone. But the third, even though he occupies the seat farthest away from her, jumps to his feet. "Would you like to sit down?"

Factors such as life satisfaction, optimism, frequent positive emotions, and infrequent negative ones are known to promote health and longevity.[38] It is less well known that altruistic behavior, for example volunteering to work for a charitable organization or giving up your seat to someone who needs it more than you do, not only improves psychological well-being but also physical health and longevity.[39] By contrast, research has found that people who exhibit high levels of hostility show signs of faster cell aging[40] and face greater risk of developing coronary heart disease and of dying from all causes.[41] Of course, the relationship works the other way around, too: being happy and healthy makes it more likely that people will be able to help others. But this is an example of that rare beast, the virtuous cycle, because it seems that being happy makes you feel more altruistic, and being altruistic makes you feel happier. So, all other things being equal, folks such as the young man who gave up his seat for Molly in the waiting room are more likely to lead longer, healthier, and happier lives than people who exhibit a similar attitude to the men who ignored her. It would seem his reward will not only be in heaven, if such a wonderful place exists, but also right here on earth. A Buddhist would say it's simply kamma in action.

The touching scene that played out in the waiting room did actually happen, but as you may have suspected, things were not all they seemed. For one thing, the setting was not a dentist's or doctor's practice but the waiting area for several labs in the psychology department at Northeastern University in Boston, Massachusetts. For another, the woman on crutches made her dramatic entry no fewer than fifty-six times. Sometimes she got lucky and was offered a seat, sometimes she wasn't. Fortunately, she was only acting the part, as was the man fiddling with his mobile phone, who was not updating his Facebook status or sending a text message but starting a timer. In fact, the only person who wasn't a stooge was the man or woman sitting farthest from Molly. This person had no idea they were part of an experiment designed to investigate whether a short course of mindfulness meditation can make people more compassionate.[42] Participants had been randomly assigned either to take a three-week mindfulness meditation course or a three-week mental training program involving games designed to improve memory, attention, and problem-solving skills. Both courses were administered via a smartphone app and required about the same time commitment. The results were clear-cut. Among those who had taken the mental training course, 16 percent gave up their seat to the woman on crutches, whereas 37 percent of those who had taken the mindfulness meditation course gave up their seat to her. Men were as likely as women to give up their seat. By seating two stooges between the subject and Molly, the scientists maximized the "bystander effect"—the well-known tendency for people to ignore a stranger in need if they perceive that someone else is better placed to help—so the altruistic bar was set particularly high.

Some forms of Buddhist meditation are specifically aimed at cultivating wholesome mental qualities known as the Four Immeasurables: loving-kindness (wishing oneself and others happiness), compassion (the desire to alleviate one's own or another's suffering), empathetic joy (rejoicing in the success and happiness of other people), and equanimity (being neither apathetic nor overexcited, meeting triumph and disaster with calm acceptance). Buddhists believe the Four Immeasurables to be essential for psychological health—reflecting their conviction that all living creatures rather than being separate, independent entities are intrinsically connected—which is why they spend a lot of time trying to foster these qualities through meditation. In Mahayana Buddhism, compassion is considered the ultimate source of happiness. So perhaps the most surprising feature of the research described above was that altruism, kindness, or compassion did not get a single mention during the three-week meditation course, which was simply designed to improve mindfulness. The explanation may be that it is impossible to maintain a state of mindfulness (present-focused, *nonjudgmental* awareness) in the absence of equanimity, kindness, and compassion—both for oneself and others. Without the Four Immeasurables we couldn't practice acceptance, because whenever we were confronted by a difficult emotion or thought, we would get caught up in it and be swept away by rumination and evaluation based on our various prejudices and preconceptions.[43]

"Mindfulness has compassion at its heart because what you're asking people to do is become aware of negative thoughts and negative feelings and to meet them with curiosity and, crucially, with kindness," said Willem Kuyken, director of the Oxford Mindfulness Centre, when I asked him why compassion wasn't an explicit part of mindfulness-based cognitive therapy (MBCT) for people at risk of depression. He believes teaching it explicitly could prove counterproductive. "If you invite someone to be self-compassionate who has a very strong inner critic and very strong negative thinking, they can have a real kickback against that. Teaching it implicitly is a more powerful way of doing exactly the same work." Regardless of whether it is taught or merely implied, research by Kuyken and his colleagues suggests that increased self-compassion plays a vital role in the success of mindfulness therapy for depression.[44] It has also proved indispensable to the success of mindfulness-based relapse prevention for drug addicts. Sarah Bowen from the University of Washington, who developed and teaches the program, explained to me why she believes compassion and forgiveness are so important for sustained recovery from an addiction. "It's about saying, 'I'm human and it's hard to be human. But I'm doing my best.' We encourage our clients to forgive themselves for ways they have hurt people, forgive other people for ways they have hurt them, and forgive themselves for ways they have hurt themselves. So many of us in the West have a hard time with self-esteem but these guys especially. There's so much shame and they oftentimes see themselves as utter failures, so sending kindness to themselves can be very difficult. We work with that. Gently, very gently."

We could all use a bit more self-compassion. There are meditation practices designed specifically for this purpose—you will find a typical compassion/loving-kindness meditation to try for yourself at the end of this chapter. The basic idea is first to think about one's nearest and dearest and then to cultivate the wish that they be free from suffering, pervading the mind with that feeling of compassion, then gradually extend this to people you know less well and finally to all living beings. Research into the effects of this kind of meditation on the brain is at an early stage, but preliminary studies suggest that in expert meditators, the practice activates circuits involved in emotion and empathy, including the insula and amygdala.[45] But some neuroscientists who study meditation are beginning to think that, at least where compassion is concerned, their focus on the brain is far too narrow. Richard Davidson tells a story about his first visit to the Dalai Lama's residence in Dharamsala, India, with colleagues from the University of Wisconsin in 1992. The scientists were keen to show off their work, so they set up a demonstration for the benefit of a couple of hundred young monks to show how they investigated compassion in the lab. Neuroscientist Francisco Varela, who cofounded the Mind and Life Institute with the Dalai Lama in 1987, sat at the front of the audience room, while Davidson and his team stood between Varela and the monks, hooking him up to the monitoring equipment and shielding him from view. When everything was ready and they stepped aside to reveal the scientist wearing an electrode cap, the young monks burst out laughing. The scientists, their pride slightly wounded, would later discover that it wasn't the sight of Varela's head plastered with electrodes that they found so hilarious but the idea that these Western scientists thought they could study compassion by monitoring electrical activity inside someone's head rather than their heart.[46] Two decades would pass before science addressed the monks' conviction that their heart played a crucial role in compassion, but a few years ago Davidson and his colleagues discovered that heart rate increases during this meditation, and significantly more among expert meditators than among novices. They found that this increased heart rate is tightly coupled with increased activation of the insula, which as we've seen is a part of the cortex intimately involved in the conscious experience of internal bodily states.[47]

The most comprehensive study of compassion training to date looks likely to show that it has the potential to increase altruism and other prosocial behaviors. At the time of writing, this research is still in press, but I had the good fortune to be in the audience when an effervescent German scientist named Tania Singer bounced onto the platform at the 2014 international symposium of the Mind and Life Institute in Boston to reveal some of the results. She said her team at the Max Planck Institute for Human Cognitive and Brain Sciences in Leipzig, Germany, had compared three types of meditation training—for cultivating mindfulness, empathy, and compassion—in a large study that lasted a year. All participants took the three training modules but in a varied sequence. Each type of meditation led to distinctive changes in their brain and their behavior. For example, empathy training improved participants' performance on tasks that tested theory of mind (the ability to intuit the mental states of others), whereas compassion training had no effect on this measure but was the most effective at increasing participants' generosity and other altruistic behaviors. Remarkably, it also reduced the time it took their bodies to recover after an acute stress challenge, such as a harsh interview or a scary encounter with a virtual-reality spider.

If results such as these are confirmed, the implications are profound. In the research described at the start of this section, just twelve minutes of mindfulness training a day delivered via a smartphone app made people significantly more likely to give up their seat for a stranger in pain. Reporting their results in the dry, cautious language of science, the researchers rounded off their paper with a surprisingly Utopian vision.[48] They wrote that their findings "point to the potential scalability of meditation as a technique for building a more compassionate society. . . . As our past work has shown, grateful beneficiaries of aid evidence a marked increase in their likelihood to subsequently extend help to others, even if these others are complete strangers. Accordingly, the potential for efficient, fairly rapid deployment of mindfulness-based benefits on compassionate responding appears worthy of increased investigation."

Over the past century, medical science has taken giant steps in developing treatments for mental illness, but there has been far less research into the potential benefits of cultivating positive human qualities such as equanimity, compassion, and altruism—which is surprising, given the vital role they play in creating a harmonious society. The Buddha certainly set a very high value on compassion. One story describes how he was conducting an inspection of the dwellings in a monastery with his attendant Ananda when they came across a monk who was so weak with dysentery he had fouled himself and was lying helpless on the floor.[49] The Buddha asked the sick man why no one was taking care of him. He replied, "I don't do anything for the monks, lord, which is why they don't attend to me." Instead of ordering someone else to do it, Ananda and the Buddha washed him and carried him to a bed. The Buddha then called all the monks together and told them, "Monks, you have no mother, you have no father who might tend to you. If you don't tend to one another, who then will tend to you? Whoever would tend to me should tend to the sick."

Guided Meditation: The Warm Glow

Many years ago, when I was a regular churchgoer, I would sometimes be asked to lead the congregation in prayers of intercession. As I prepared what I was going to say on the morning of the service, it was always a huge challenge to avoid making these requests for divine intervention sound like a shopping list. I would usually end up asking God to take care of the queen, the government, the sick, the disadvantaged, the downtrodden, to sort out peace in the Middle East and visit the households on a particular local street whose turn it was for special attention. In the end, what made this exercise impossible for me was that I was beginning to doubt the power of prayer to change anything. My days as a Christian were numbered.

I was wrong about the power of prayer. Prayers of intercession and compassion meditation have a lot in common, as you will see shortly, and certainly the latter has been found by scientific research to improve the meditator's mood, reduce stress, and make them less angry, kinder, more generous, more altruistic people.[50, 51] Unlike Christians, Buddhists don't expect to change *other* people's lives as a direct result of their meditations, but it seems that simply repeating inwardly the heartfelt desire that one's fellow humans be happy and free from suffering has a powerful effect on the person expressing such wishes.

It's worth remembering that nonjudgmental mindfulness is impossible unless one is prepared to show compassion toward both oneself and others. For Buddhists, feelings of compassion and kindness are also an expression of their conviction that all living beings are interconnected, in contrast to the more intuitive belief that each of us is a distinct Self, an island separated by deep water from everyone else's island.

The exercise below combines traditional loving-kindness and compassion meditations, and like all the others in this book should be treated as a framework for practice rather than a rigid set of instructions. During this meditation, you start by extending feelings of compassion and kindness toward a loved one, then progressively expand these to encompass yourself; a comparative stranger; someone with whom you have a difficult relationship; and finally the entire human race.

First calm the mind by focusing on your breathing for a few minutes (see "Guided Meditation: Only the Breath," page 56). When you're ready, wish someone particularly close to you well by silently reciting these words:

All desire health and happiness.
May you be free from mental suffering;
May you be free from physical suffering;
May you experience joy and ease.

Envisage this sentiment of wishing your loved one well as the unconditional love of a mother comforting her child. Nurture this warm, heartfelt sensation almost as if it were a glowing ember in the center of your chest. Breathe air into this part of your body and out again, sustaining and feeding the ember, this generous internal source of warmth and compassion.

Recall a time when he or she has suffered, and as you repeat the words silently, imagine the glowing ember burning more brightly as you envelop and ease their suffering with your compassion.

Now extend these wishes toward yourself, using the same formula of words but replacing *you* with *I*. After each line, observe any thoughts and feelings that arise, without pursuing them. Call to mind a time when you have suffered, then extend the same warm glow of compassion toward yourself, the same desire to alleviate this suffering.

Do exactly the same for someone you may see regularly but don't know personally, such as a neighbor, fellow commuter, or someone who works in the same building. Imagine a time when *they* have suffered.

Now extend these wishes to someone with whom you have a difficult relationship, using the same form of words, the same desire to alleviate their suffering.

Finally, extend your well-wishing to everyone, to the whole of humanity:

All desire health and happiness.
May people everywhere be free from mental suffering;
May people everywhere be free from physical suffering;
May they experience joy and ease.

CHAPTER NINE

THE FALL

And the Lord God said unto the woman, "What is this that thou hast done?" And the woman said, "The serpent beguiled me, and I did eat."

—Genesis 3:13 (AV)

We are standing on high ground surveying the golden Tugen Hills of eastern Africa. After a cold night under a cloudless sky, the sun is starting to warm the chilly air. In the valley below lies a glittering lake, and above, the ragged fringe of an ancient forest is alive with birdsong and the calls of monkeys. Scan the trees and you might just make out a dark shape resembling an adult male chimp standing tall and motionless on a horizontal bough twenty feet above the ground, his left hand curled around a branch above his head. The ape gazes far out over the valley, which is dotted with roaming elephants and grazing impala, then scrutinizes the dangerous hinterland between the trees and the savannah, searching for predators. Satisfied, he lays the palm and fingers of his right hand over the dome of his head from ear to ear like a cap. Immediately half a dozen kin emerge into the open from the protecting shadow of the wood. The lookout runs along the branch and then uses his long, powerful arms to lower himself swiftly to the ground. Walking on their knuckles, the apes shoulder their way through the tall grass, heading in the direction of the lake. Soon only a slight disturbance of the grass betrays their progress, but once in a while one stands up to scan the surrounding terrain, walks a few paces, then disappears from view again.

A scene like the one we have just imagined may have played out six million years ago in the Tugen Hills of what is now central Kenya. Those chimp-like creatures may have been our own forebears, members of a species called *Orrorin tugenensis* that had only recently descended from the last common ancestor of chimpanzees and humans, whose lineages split some six to seven million years ago in Africa. The ancestral chimps stayed behind in the relative safety of the forests, while the bolder hominins climbed down from the trees and ventured out into the open during daylight hours to hunt and forage.

What have these apes got to do with mindfulness? I introduce them to review potential solutions to a long-standing mystery about mental illness, one that may have some bearing on the protective effect of meditation: Why are mental illnesses such as schizophrenia, bipolar disorder, anxiety, and depression so common in populations the world over? Each of these conditions has a significant genetic component, and their symptoms manifest early in life. We know that they not only raise mortality rates and reduce one's chances of having children, but also the life expectancy and fertility of those offspring. Natural selection is usually ruthlessly efficient at weeding out genes that compromise an organism's chances of survival and reproduction, and it makes no exception for us. Evolutionary theory therefore predicts that genes like the ones that predispose people to develop mental illness are less likely to be passed to subsequent generations and will quickly disappear. So why are mental illnesses still with us?

It is easy enough to explain the persistence of rare genes behind debilitating disorders such as Huntington's disease that strike when people are approaching middle age—long after they have had children and unwittingly passed them to the next generation—but mental illnesses are usually diagnosed in our teenage years, twenties or early thirties, in other words, during the peak reproductive years. There can be little doubt that these conditions diminish patients' reproductive fitness. Research suggests that the most widespread psychiatric conditions have a moderate but significant effect on fertility, whereas the rarer disorders have a very large effect. The lifetime prevalence of anxiety disorders, for example, is around 30 percent, their average age of onset is eleven, and the fertility of people with these conditions is 10 percent less than the average for the total population. At the other end of the scale, schizophrenia is diagnosed in just under one in every hundred people, its signs generally come to the notice of doctors when patients are in their early twenties, and fertility is reduced by 60 percent on average. Estimates of the lifetime incidence of major depression vary enormously, from 20 percent to 50 percent. Average age of onset is somewhere between the midtwenties and early thirties, and the condition reduces fertility by about 10 percent overall.[1-4]

The persistence of mental illnesses in human populations could easily be explained away if they were caused by some widespread environmental factor, such as poor diet, overcrowding or pollution, but they have a substantial genetic component. Studies that compare identical and nonidentical twins estimate that for bipolar depression and schizophrenia, around 80 percent of the variation in the risk of developing these illnesses can be explained by genes and only 20 percent by the environment. For depression and anxiety disorders, around 40 percent and 30 percent respectively of the variation in risk is down to genetic factors.[5] So far, however, despite sweeping the genomes of tens of thousands of people, scientists have collared only a handful of genes that increase the likelihood of any mental illness, and each one accounts for only a fraction of the overall risk.[6, 7] This suggests that common mental illnesses such as depression and anxiety are caused by a complex interaction between lots of very common genes (each with a tiny individual effect) and environmental influences.

As you can imagine, the persistence of highly heritable mental illnesses that strike in early adulthood—not only reducing patients' fertility but also that of their offspring—has had psychiatrists scratching their heads. One explanation is that while the many gene variants responsible harm the reproductive chances of patients *themselves*, in the lower doses found among their siblings they may actually increase fertility. In genetics this is known as "balancing selection." In medicine, the classic example is the gene for sickle cell anemia: if you inherit a copy of this gene from only one of your parents, it will protect you from infection by the malaria parasite, but if you inherit one copy from each parent, your red blood cells twist into a rigid sickle shape, causing severe pain and increasing the risk of stroke. This explains why the sickle cell gene and the disease it causes persist in sub-Saharan Africa, where malaria is rife.

A variant of balancing selection known as the "cliff edge" hypothesis proposes that a particular genetic trait is beneficial in reproductive terms but only up to a point, beyond which it is harmful. One beneficial trait that has long been associated with mental illness is creativity. According to the cliff edge hypothesis, a little extra creativity as a result of being dealt a particular genetic hand increases one's reproductive success, whereas too much of the same trait is deleterious. The roll call of famous writers, performers, and artists who have been afflicted by mental illness is certainly very long. Aristotle claimed that "no great genius has ever existed without a strain of madness"[8] and there is now evidence that the ancient Greek philosopher—no creative slouch himself—was on to something. A study of 300,000 people with severe mental disorders in Sweden found that those with bipolar disorder and the healthy siblings of patients with either bipolar or schizophrenia were more likely to be involved in creative professions (defined in this study as having an artistic or scientific occupation).[9] Could one of the "cliff edge" benefits of enhanced creativity be increased reproductive success? If so, in siblings this would compensate for the reduced fertility of patients themselves and help explain why the genes responsible for mental illness are so persistent. There is some evidence this may be the case. A questionnaire-based study of 425 adults in Britain asked them about their creative output and the number of steady relationships and sexual partners they had had since they were eighteen. It also measured "schizotypy"—four personality dimensions that are characteristic of people who have been diagnosed as having a mental illness but are also found in varying degrees throughout the population: unusual perceptual and cognitive experiences such as hallucinations and magical thinking; difficulties with attention and concentration; lack of enjoyment and social withdrawal; and nonconformity, including unstable moods and behavior, disregard of rules, and violence or recklessness. People who scored high on unusual experiences were more likely to be poets or artists, and this in turn appeared to increase their mating success. Nonconformity was also associated with greater mating success. Lack of enjoyment and social withdrawal, however, tended to decrease both creative activity and mating success.[10]

So the unusual experiences often associated with mental illness could make people more creative, which in turn works as a uniquely human form of mating display to increase reproductive success. But there may be a limit to these benefits beyond which the experiences have a profoundly negative effect—it tips fertility over a cliff edge. A study that scoured the medical records of 2.3 million people living in Sweden, accumulated over a twenty-year period, found that patients diagnosed with a wide range of mental illnesses had significantly fewer children than average. However, there was also some balancing selection, because the siblings of patients with depression and substance abuse disorders had significantly *more* children than average, which in the case of depression was more than enough to make up for the reduced fecundity of their less fortunate brothers and sisters.[11] Nevertheless, "gene-environment" interactions may also play an important role. For example, you may benefit from being highly creative if you were born into a wealthy, supportive family, but there could be negative consequences if you were born into less favorable circumstances.

We are still a long way from finding a definitive solution to the puzzle of why mental illnesses remain so prevalent. Creativity is almost certainly only a small part of the story: the figures suggest that other "cliff edge" factors must also be contributing. Language ability may be an important one, as I will explain shortly. But it seems likely that the large number of common genes that make people vulnerable to anxiety and depression—and even more so if they experienced trauma, abuse, or deprivation in childhood or face adverse conditions during adulthood—are preserved because they confer advantages in smaller doses and under more favorable circumstances. The picture is a little more complicated for mental illnesses that have the biggest effect on fertility, such as bipolar disorder and schizophrenia. The Swedish researchers found that the sisters of patients with these diagnoses did have more children than average, but this boost was not enough to compensate for their siblings' reduced mating success. In contrast to depression and anxiety, these rarer illnesses may persist not only through balancing selection but also because they are partly caused by random mutations in parents' or grandparents' DNA—particularly in the sperm of older fathers. There is good evidence that the children of older men have a higher risk of developing schizophrenia and bipolar disorder. It is worth remembering, though, that even in these conditions there is still plenty of scope for gene-environment interactions that make them more or less likely to develop.[12]

In the very first sentence of this book I quoted Ajahn Amaro, the abbot of Amaravati Buddhist Monastery in the UK, who during my brief stay as a guest there in 2014 dropped into our conversation the idea that we are *all* mentally ill. He was not talking about the official diagnoses you can find by thumbing through the psychiatrists' bible, the *Diagnostic and Statistical Manual of Mental Disorders*, but our shared vulnerability to the three psychological "poisons" that Buddhists believe to be the root cause of human suffering: craving, aversion, and delusion. This isn't to deny the ample evidence that some people are more vulnerable than others to mental suffering as a result of their genes and circumstances. Rather, it focuses on our common inheritance as a species: the factors in our shared psychological makeup that predispose all of us to addiction, anxiety, depression, and even psychosis if we have the misfortune to experience difficult conditions while growing up or when faced with challenging situations later in life such as physical illness, a relationship breakup, becoming unemployed, or losing a loved one.

In tune with this Buddhist perspective, many psychiatrists have been coming to the conclusion that clear-cut diagnoses such as generalized anxiety disorder, depression, bipolar disorder, and schizophrenia are a convenient fiction.[13, 14] In reality, the distinctions between these conditions are blurred, and their constituent symptoms are widespread in the general population. So for example, people with severe depression often experience delusions and hallucinations, and those with an anxiety disorder are likely to have some symptoms of paranoia and be more prone to depression. Patients who have been diagnosed as having bipolar disorder or schizophrenia also have symptoms in common. As a result of this lack of clear boundaries, a single patient might be given several different diagnostic labels. At the same time, geneticists and neuroscientists have failed to find any justification for assigning people to distinct categories, because the overlap between them in terms of genetics, brain structure, and function is so great. It makes more sense to consider mental illness as a collection of shared symptoms that vary in severity on a continuum with everyday experiences in the general population. A study published in 2013, based on interviews conducted by the UK's Office for National Statistics and standard psychiatric questionnaires, found that 20 to 30 percent of adults continually worry that certain people are against them and might exploit or hurt them. One in six people spend a lot of time wondering whether they can trust their friends or work colleagues. About 10 percent sometimes feel others are watching them, staring at them, deliberately acting to harm them, or trying to control their thoughts. At the far end of this continuum, 1.8 percent fear there are plots afoot to cause them serious harm or injury.[15] "If we take the rates of endorsement of paranoid items in our study at face value," the researchers conclude, "they suggest that paranoia is so common as to be almost normal."

A little paranoia is a good thing, of course, because wariness about the intentions of others, particularly strangers, is prudent. If someone phoned you out of the blue claiming to be from your bank and asked for your security details, you would be wise to refuse. Paranoia only becomes a problem when it is unrelated to reality, causes distress, and interferes with normal, everyday functions. In keeping with the cliff edge hypothesis, to exercise a little caution around other people is adaptive, whereas paranoid delusions about them may lead to social isolation and plunge the most vulnerable into mental illness. The same is true of those other two poisons of the mind singled out by Buddhism, craving and aversion. Appetites of all kinds are a biological necessity, but in excess they lead to clogged arteries, dependency, and addiction. Anger and aversion might once have made all the difference in a primitive fight for survival, but they are counterproductive in the modern world, where compassion, compromise, and cooperation are more beneficial for everybody concerned.

Thinking about Ajahn Amaro's words, I was reminded of a conversation between Alice and the Cheshire Cat in *Alice in Wonderland*:[16]

"What sort of people live about here?"

"In that direction," the Cat said, waving the right paw 'round, "lives a Hatter; and in that direction," waving the other paw, "lives a March Hare. Visit either you like: they're both mad."

"But I don't want to go among mad people," Alice remarked.

"Oh you can't help that," said the Cat; "we're all mad here. I'm mad. You're mad."

"How do you know I'm mad?" said Alice.

"You must be," said the Cat, "or you wouldn't have come here."

You probably remember the scene that follows. Alice chances upon the Mad Hatter, March Hare, and Dormouse having a tea party. When she joins them, the Hatter and March Hare gleefully expose the logical flaws in everything she says. Their creator, the Victorian mathematician Lewis Carroll, had the kind of analytical, scientific mind that relished this sort of exercise. Like the Hatter and March Hare, we too have the capacity for unassailable logic, but like them we are prone to "moments of madness" when we do something aggressive, irrational, and perhaps even a little silly—like trying to stuff a sleepy dormouse into a teapot.

It seems mental illnesses are not triggered by a smattering of rare genes carried by an unfortunate few. If they were, the conditions would be less widespread and we would probably have developed a set of diagnostic genetic tests by now. Rather, they are caused by a multitude of common genes interacting in complex ways with each other and the environment. These genes are either recent mutations or are maintained by balancing selection. No other animal species, even the smart ones, suffers from debilitating mental illnesses. They certainly don't commit suicide (sacrificing oneself for the colony or for one's offspring isn't the same thing at all). At what point in our evolution did things start to go so wrong? Consider a hunter-gatherer on the hostile African savannah. According to Herbert Benson, the Boston cardiologist we met in the second chapter, who discovered the relaxation response (the physiological flip side of the fight-or-flight response), our vulnerability to stress arises from the way the human mind strings out thoughts about past or future threats. "It's not just having a saber-toothed tiger in front of you, it's the *thought* of the saber-toothed tiger," he told me. Other prey animals fear predators as much as our hunter-gathering ancestors must have, and run for their lives when under threat, but they don't obsess about predators when they are no longer there or if they have already made a kill. Impala don't suffer from sleepless nights or gnaw their hooves the way we humans chew our nails. When the danger has passed they quickly go back to eating the grass. It's hard to tell, but they do seem the happier for it.

Nevertheless, there remain questions about how and why our species became such inveterate worriers and brooders, not to mention our tendency for paranoid delusions. To explain these behaviors, we may have to travel back in time some six million years to the chimp-like creatures I introduced at the start of this chapter. Fossil remains of *Orrorin tugenensis* were discovered in the Tugen Hills in 2000 and may be the earliest evidence of bipedalism in a human ancestor (another, weaker contender for this title is *Sahelanthropus tchadensis*, whose skull was found in Toros-Menalla in Chad). The remains include pieces of femur that suggest *Orrorin* was adapted for walking upright on two legs, while other features indicate it still spent much of its time up trees. We will probably never know for certain whether *Orrorin* is an ancestor of Australopithecines, the hominins that preceded the appearance of our own genus, *Homo*, about 1.8 million years ago, though some researchers claim that the bones point that way.[17, 18] Regardless of any difficulty reconstructing the exact evolutionary route our ancestors took when they stood up and walked on their hind legs, it would eventually have profound repercussions.

The immediate advantage for apes on foraging expeditions in the open may have been the ability to see where they were going in long grass and spot predators from a distance. On the rare occasions when modern chimps venture out into the savannah, they sometimes stand up on two legs to get a better view. A variety of advantages of walking upright all the time have been proposed. These include the ability to travel great distances efficiently; prevent overheating by exposing a smaller surface area of the body to the midday sun; and freeing the hands to carry provisions, tools, and crude weapons such as stones and spears—and throw these projectiles with deadly accuracy.[19] But perhaps the most far-reaching advantage—the one that would eventually lead to the extraordinary abilities that define our species—was to free the hands for communication. As I will explain shortly, we may have paid a high price for this breakthrough.

According to psychologist Michael Corballis, who studies the evolution of language at the University of Auckland in New Zealand, crude hand signals like the one I envisaged being used by the lookout ape to signal safety at the start of this chapter evolved over the course of millions of years into a sophisticated sign language that would one day lead to the emergence of vocal language. Corballis and others have argued strongly against the idea that spoken language appeared almost out of nowhere in an isolated "big bang" moment within the past 100,000 years as a result of a chance genetic mutation in a single individual.[20] Evolution tends to inch forward in tiny, incremental steps rather than in huge leaps. Changes build one on top of another over eons like a stalagmite growing on the floor of a cave splashed by countless drops of water. Corballis believes that the neural hardware required for grammatical language evolved over millions of years in the service of an increasingly sophisticated communication system involving hand gestures and facial expressions, perhaps supplemented with oral clicks like those still used by the Hadza and San groups in Africa. He cites several lines of supporting evidence, including the fact that our cousins the chimpanzees and bonobos can be trained to communicate using abstract symbols. Primatologist Sue Savage-Rumbaugh successfully taught a bonobo called Kanzi to use 256 symbols on a keyboard denoting objects and actions, and the ape himself devised several hand gestures to supplement them.[21] Even in the wild, chimpanzees have a repertoire of manual gestures that they can use flexibly to communicate depending on the context.[22] There is no evidence that nonhuman apes can learn grammatical rules in order to string together words to make complex sentences, but the ability of our nearest relatives to recognize and use symbols suggests that our common ancestor had the neural wherewithal to develop a simple symbolic language that became increasingly sophisticated over time. Corballis points out that modern sign languages are now recognized as having all the complexity and scope of speech.

There are also clues in our brains that spoken language evolved from sign language. In the 1990s, neuroscientists at the University of Parma in Italy discovered a type of nerve cell in the brains of macaques that they called mirror neurons, because they not only fired when a monkey made an intentional movement of its hand but also when it saw one of its fellows or even a human make the same movement.[23] These mirror neurons are in the monkey's frontal lobe at a spot that corresponds precisely to a region in the left hemisphere of the human brain known as Broca's area, which is not only closely involved in speech but also in controlling complex hand movements. This has led Corballis to suggest that at some point during our evolution, vocal language was incorporated into a system that had already evolved for communication using manual gestures. It is at this point that the evolutionary stories of mental illness and language come together, because mirror neurons are now considered part of a more extensive mirror system in the human brain that overlaps with much of the default mode network—the constellation of regions that light up whenever we are not engaged in an external task.[24] As we have seen in previous chapters, the network is intimately involved when we reflect about our place in the social world, our past experiences, and our future plans, but it has also been implicated in a range of mental illnesses. The default network is the brain's simulator: it not only creates our sense of self but also builds a representation of other people's minds—a "mirror" version of their world—allowing us to see things from their perspective and intuit their beliefs, thoughts, and intentions ("theory of mind"). And of course the default network is also the mental time machine we use to rewind into the past and fast-forward into the future.

Mental time travel and theory of mind are indispensable for complex language. Whether speaking or signing, language allows us to pass on information about things that have already happened or things that might happen. A mutual understanding of each other's perspectives—what the other person knows and what they don't know, and your shared beliefs—further enhances the effectiveness of communication. The subtleties of metaphor, irony, and sarcasm, which work only if at least some of the people with whom you're communicating know not to take your words literally, would be impossible without theory of mind. It therefore makes sense to envisage language and the default mode developing in lockstep over the course of human evolution as our mental simulations became increasingly powerful, allowing us to muse about the past, speculate about the future, and infer the mental states of others. So language became the very stuff of thought. We were now talking to ourselves, providing a running internal commentary of everything that happened and might happen. Perhaps unsurprisingly, there is ample evidence that the default mode, which facilitates language, is also involved in mental illness. People with depression appear to have trouble suppressing the default mode in order to focus on the task at hand.[25] This manifests as a tendency to slip into rumination or brooding—in other words, excessive negative "self-referential thought" that can interfere with everyday activities.[26] Research suggests that even patients in remission from depression still struggle to control the default mode, which may explain their greater cognitive reactivity—the way small bursts of sadness that other people shrug off instead trigger the streams of negative thought that can lead to a relapse.[27] Other studies have implicated a failure to regulate the default mode in anxiety disorders and attention deficit hyperactivity disorder.[28, 29] Finally, there is solid evidence that control of the network is impaired in patients with schizophrenia and bipolar disorder.[30–32] Because of the default mode's central role in self-referential thinking and inferring the mental states of others, when it runs out of control this may blur the distinction between internal thoughts and external stimuli, triggering hallucinations and promoting paranoia about the intentions of others.

The link between the default mode, language, and mental illness is a dangerously intimate one. In the study of creativity in Sweden described above, accountants were no more likely than controls—and in some cases even less likely—to suffer from the mental illnesses investigated, whereas authors were twice as likely as controls to have schizophrenia or bipolar disorder. They were also more likely to be diagnosed with depression, anxiety disorders, alcohol and drug abuse, and to take their own lives.[33] A follow-up study that analyzed the records of more than a million Swedish people confirmed these findings.[34] So it seems language is a dangerous mistress. A love of numbers might make for a happier life.

The affair may have started innocently enough with perhaps a few dozen manual gestures employed by our ape forebears to exchange information among themselves some six million years ago. Things only began to get out of hand comparatively recently, but all along, the evolutionary impetus for these improvements in communication was social cohesion. Evolutionary psychologist Robin Dunbar has discovered that the number of individuals in primate social groups increases in direct proportion to the relative size of the orbitofrontal cortex (so called because it is located immediately above the orbits of the eye sockets; see Figure 6, page 211).[35] This extends into the ventromedial prefrontal cortex, the area we first encountered in chapter 5, "The Man Who Disappeared," as an integral part of the brain's "Self app," which not only simulates our sense of having a solid, unchanging self but also other people's perspectives. Dunbar found that the relative size of a primate's orbitofrontal cortex predicts the sophistication of its theory of mind or "mentalizing" abilities—in other words, how good it is at inferring

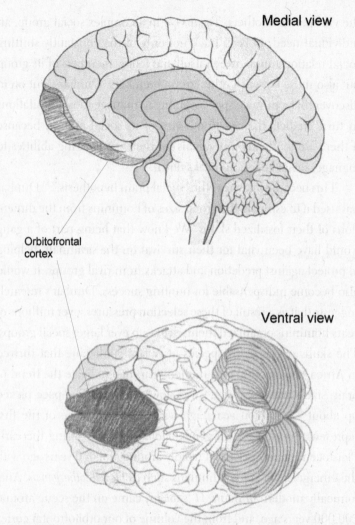

Medial view

Orbitofrontal
cortex

Ventral view

Figure 6. Human evolution. The orbitofrontal cortex (which extends into the ventromedial prefrontal cortex on the inner surface of each hemisphere) is essential for mentalizing or theory of mind, inferring the motivations and mental states of others. Among primates, mentalizing helps to maintain social cohesion, which may explain why ape species with larger orbitofrontal cortices have larger maximum social group sizes. This part of the brain expanded enormously in the course of hominin evolution as our ancestors banded together in ever larger groups.

the viewpoints of others. To survive in a complex social group, an individual needs to keep track not only of the constantly shifting social relationships between itself and fellow members of its group but also those between other group members. Dunbar went on to discover that a primate species' ability to manage these simulations in turn predicts the maximum size of its social groups, because if there are too many individuals for their mentalizing abilities to manage, social cohesion breaks down.

This became known as the "social brain hypothesis."[36] Dunbar has used it to estimate the group sizes of hominins from the dimensions of their fossilized skulls. We know that being part of a gang would have been vital for their survival on the savannah, helping to protect against predation and attacks from rival groups. It would also become indispensable for hunting success. Dunbar's research suggests that as a result of these selection pressures, over millions of years hominins organized themselves into ever larger social groups. The skulls of the many species of Australopithecine that thrived in Africa from around four million years ago indicate the trend in brain size—and group size—was slowly upward. The pace picked up about 1.8 million years ago with the appearance of the first representative of our own genus, *Homo ergaster*, during the early Pleistocene, and accelerated rapidly around 500,000 years ago with the emergence of archaic humans such as *H. heidelbergensis*. Anatomically modern humans, *H. sapiens*, came on the scene around 200,000 years ago, and from the volume of our orbitofrontal cortex (and the mentalizing abilities it facilitates), Dunbar predicts that the maximum group size of our species is around 150 people.

The challenge for evolutionary psychologists like Dunbar is to explain how hominins found the time and energy to hold their increasingly large social groups together. Most primates keep the peace and ensure social cohesion through mutual grooming, but this is time-consuming and so there is a limit to the number of individuals one can groom while still having enough time during daylight hours for traveling and foraging. Dunbar proposes that our hominin ancestors first maintained cohesion in their burgeoning groups by making extra time for grooming through other savings, such as anatomical adaptations for more efficient long-distance walking and changes in diet to more energy-rich foods, including tubers. Much later, he says, there would be singing, dancing, and laughter; better tools and weapons; more meat, fire, and cooking. But vocal language would eventually become the most effective social glue of all. Estimates of when speech emerged vary enormously, at anything from 500,000 to 50,000 years ago. But finally there would come storytelling, gossip, rhetoric, and religion—the mixed blessings of spoken language that still hold much of the modern social world together.

As we have seen, language, mental time travel, sophisticated mentalizing abilities, and self-referential thought come as a package courtesy of the brain's default mode network. The organization of this network in humans sets us apart from other apes, making possible impressive feats of creativity, forward planning, and problem solving. It glues our unusually large social groups together by allowing us to communicate and infer the mental states of others to a degree unprecedented among primates. Having a highly sensitive theory of mind has some odd effects, though—some entertaining and others less so. Our minds are tuned to see the semblance of human faces everywhere we look, even in punctuation ;), pieces of toast, and random cloud formations. In a house where I lived as a child, every night I saw the devil leering down at me from the curtains of my bedroom, though in the morning all I could see were flowers and foliage. We have a tendency to infer conscious agency where there is none. A puppeteer I sometimes watch performing in Leicester Square in central London makes a grinning skeleton puppet dance to recordings of popular songs. A particularly poignant favorite is Bobby McFerrin's "Don't Worry, Be Happy." He stands over the skeleton pulling its strings in plain sight of the spectators, but his skill is such that one is drawn into the amusing but slightly creepy illusion. Why else would we stop and watch, spellbound? This is what an overactive theory of mind will do for an ape. It's little wonder that wherever our ancestors looked they saw ghosts, spirits, and gods. We like to think we are above such delusions, but if you have ever worked in an office, you will have heard colleagues curse their computers as if a ghost in the machine were deliberately trying to thwart them. You have probably done some of the cursing yourself. This isn't considered unusual behavior, but under the extreme stress of situations such as sleep deprivation or solitary confinement, people routinely start to experience the kind of delusions and hallucinations popularly supposed to be the exclusive preserve of the mentally ill.[37, 38]

As we saw in chapter 5, "The Man Who Disappeared," we use the same neural equipment to simulate the mental states of *others* that we use to simulate our sense of *selfhood*. So this was the evolutionary bargain our species struck in exchange for the benefits of language, mental time travel, and theory of mind. The Self was born on the African savannah. It seems humans intuited long ago that something had gone awry in their distant past, and that it had something to do with the origins of self-consciousness. They tried to make sense of it through the legends and religious stories they told themselves. Most famously, the Abrahamic religions traced the origins of our predicament to the Garden of Eden, where

Adam and Eve were said to have eaten forbidden fruit from the tree of knowledge—"and the eyes of them both were opened, and they knew that they were naked."[39] They concluded that humanity had been much happier before this abrupt fall from grace. The fourteenth-century Christian contemplative who wrote *The Cloud of Unknowing* perceptively laid the blame on the disordered mental simulations that now plagued the untrained human mind:[40]

> *Imagination is the faculty by which we form images of all absent and present things; and both it and the matter on which it works are contained by consciousness. Before humanity sinned, imagination was so obedient to reason (to which it is, as it were, a servant) that it never supplied it with any disordered image of a bodily object or any delusion of a spiritual object; but now it is not so. Unless it is controlled by the light of grace in the reason, it will never cease, asleep or awake, to form alien and disordered images of bodily objects, or else some delusion or other, which is nothing but a bodily conception of something spiritual or a spiritual conception of something bodily. This is always deceptive and false, and associated with error.*
>
> *In those who are newly turned from the world to devotions, this disobedience of the imagination can be clearly recognised when they are praying. For until the time comes when the imagination is to a great extent controlled by the light of grace in the reason—as it is by continual meditation on spiritual matters, such as their own sinfulness, the Passion, the kindness of our Lord God, and many other similar topics—they are quite unable to*

set aside the strange and alien thoughts, delusions and images that are supplied and imprinted in their minds by the light and ingenuity of imagination. All this disobedience is the punishment for original sin.

An alternative explanation can be found in the distant evolutionary past of the human lineage. The Tugen Hills of Kenya, where the remains of *Orrorin* were uncovered, and Lake Turkana in Tanzania, where several hominin species including *Homo ergaster* have been found, don't hold the same pull for the world's tourists as the Great Pyramids or Stonehenge—or for that matter Bodh Gaya or Bethlehem. But the sites where the petrified bones of our ancestors were unearthed act as bookmarks for some fascinating chapters in the story of our species. I believe they speak to who we are just as powerfully as do those great ancient monuments to human ingenuity and creativity.

Our evolution into such highly sociable, talkative creatures is what set us apart from the other apes and our distant hominin ancestors, allowing us to live harmoniously in large, complex social groups, but it also leaves us vulnerable to the dangers of disengaging from the reality of the present moment and succumbing to self-referential thoughts, rumination, worry, and paranoia. It is difficult to escape the conclusion that we have evolved into an ape that takes things personally. If our brains were computers, we would demand an upgrade of the operating system. As we'll explore in the next chapter, it seems that meditation—though not a quick fix—might help to "debug" the human mind, correcting the flaws that have accumulated in the course of its evolution.

CHAPTER TEN

WONDERFUL AND
MARVELOUS

*Make it thy business to know thyself, which is the most
difficult lesson in the world.*
　　　　—Miguel de Cervantes, *Don Quixote de la Mancha*,
　　　　　　　　　　　　　translated by Peter Motteux

At one time the Buddha was living at a monastery in Jeta Grove
near the city of Sāvatthī on the Ganges Plain. After returning from
their alms rounds and eating their meal, a number of the monks
were chatting in the assembly hall when their conversation turned
to the miraculous things they had heard about him, such as how
he seemed to know everything about the buddhas of old, their
names, their clans, the depth of their concentration during meditation, which ones attained the deathless state of nibbāna/nirvana.
Ananda, the Buddha's faithful attendant, was about to tell them

what he had heard: "Friends, Tathāgatas [perfectly enlightened, selfless beings] have wonderful and marvelous qualities . . ." when the Buddha himself walked in and took his seat. Ananda fell silent.

"What were you talking about?" the Buddha inquired.

"We were discussing your wonderful and marvelous qualities, sir," they replied rather sheepishly.

"Carry on, Ananda," said the Buddha, "tell them about the Tathāgata's wonderful and marvelous qualities."

Ananda cleared his throat. "Well, I have heard it said that when the Blessed One descended from heaven into his mother's womb a great immeasurable light surpassing the splendor of the gods appeared so that even the abysmal places of utter darkness, where the light from the moon and the sun had never reached, became brilliantly lit, and the ten-thousandfold world system shook and quaked and trembled. I have heard this is one of the wonderful and marvelous qualities of the Blessed One . . ."

"Go on . . ."

"Ahem . . . and the Blessed One's mother became perfect in every virtue, no kind of sickness afflicted her during pregnancy, she was blissful and free from bodily fatigue. She carried him for ten months rather than the usual nine. She gave birth standing up. Four devas caught him and when the Blessed One came forth from his mother's womb he was completely unsullied. Even so, two jets of water poured down from the sky, one cool and one warm, to wash the Blessed One and his mother. Lotus flowers bloomed all around as the Blessed One put his foot on the ground for the first time—and walked!"

"So would you like to hear another wonderful and marvelous quality of the Tathāgata?" the Buddha asked.

Everyone nodded enthusiastically. "Yes please!"

There followed a long pause and when he finally spoke it was with the faintest hint of a smile. "When a feeling arises in the mind of the Tathāgata, he knows this is a feeling. When a feeling abides in his mind and when a feeling fades away, he knows this is a feeling abiding and this is a feeling fading away. When a perception arises in the mind of the Tathāgata, he knows this is a perception. When a perception abides in his mind and when a perception fades away, he knows this is a perception abiding and fading away. When a thought arises in the mind of the Tathāgata, he knows this is a thought arising. When a thought abides in his mind and fades away, he knows this is a thought abiding and fading away. This too is a wonderful and marvelous quality of the Tathāgata."[1]

While the Buddhist scriptures are replete with breathless descriptions of supernatural occurrences, the Buddha continually tried to refocus his followers' attention on the insights he had gained into how to cultivate a healthy mind, which he believed to be infinitely more important than any magical tale. During my stay at the monastery, Abbot Ajahn Amaro told me the story outlined above. "What he points to here is that all that stuff about lotus flowers and devas doesn't really help you very much," said the monk. "What is really amazing, the real miracle, is that you can watch your own mind."

Psychologists call this "metacognition"—the ability to *think about thought*, objectively reviewing feelings, perceptions, ideas, and beliefs as they arise in the mind. Humans are almost certainly the only species that can do this. One of the aims of meditation is to break the habit of identifying with the contents of our stream of consciousness, clutching at them as if they defined us, and instead view them as discrete mental events that we can watch as they arise, linger for a while, and then fade. Metacognition is what allows you to make silent, nonjudgmental observations during meditation—and, more importantly, life in general—such as "There is calm" or "There is frustration," as opposed to "I am *so* frustrated right now!" This leads to an understanding of how the contents of the mind ebb and flow. When Buddhists talk about the insights that arise from mindfulness, this is one of them.

This ancient concept has close parallels in psychiatry, because one of the common factors known to predispose people to a range of mental illnesses, alongside things like adverse experiences in childhood and low self-esteem, is an inability to step back and review thoughts and feelings objectively. Failure to "self-regulate" in this way is implicated in a wide range of behavioral and psychological problems, including poor school performance, ADHD, anxiety, depression, and drug abuse.[2] Metacognition, by contrast, allows us to weigh our options and then consciously decide our behavior rather than acting "on impulse." Psychotherapies such as cognitive behavioral therapy (CBT) and mindfulness-based programs have been explicitly designed to nurture metacognitive skills. For example, the Mindfulness in Schools Project pioneered by the Oxford Mindfulness Centre in the UK includes an exercise in which teenagers are encouraged to visualize each thought or feeling as a bus drawing up alongside a bus stop. They have a choice whether to step on board, and even if they get on the bus only to discover later that it's not going their way, they can always hop off at the next stop.

There is nothing magical about metacognition, though it is certainly a wonderful and marvelous ability. In the previous chapter I introduced the idea that our ancestors ganged together in increasingly large social groups in order to protect themselves from predators and rival groups on the African savannah and hunt more effectively. Failure to do so would have meant an early death and the extinction of their genes. Over millions of years, these selective pressures appear to have changed our brains to make us the sociable animals we are today, steadily improving our mentalizing abilities and facilitating the emergence of complex language and mental time travel. We became experts at forming strategic alliances and playing the games of deception and self-promotion needed to get our own way in a large group of individuals with competing needs. But there was another selective pressure operating, because to rub along harmoniously with our fellow apes and prevent these larger groups from fracturing, it would also become essential to develop a measure of self-restraint: the wisdom to know when to back down in a conflict or sacrifice selfish needs for those of the group as a whole. We learned how to be patient and delay our own gratification for another day. This conscious self-restraint would have been impossible without metacognition.

Scientists are still figuring out how the human brain performs this impressive feat of introspection, but the chances are that as a relatively new addition to our mental toolbox, metacognition is facilitated by some of the most recently evolved parts of the prefrontal cortex. The brains of highly experienced meditators may offer some clues, because observing the contents of the mind is fundamental to meditation practice. We know that the brain is plastic—regions that are involved in acquiring a new skill change over time, rather like a muscle bulking up with repeated use—so areas essential for metacognition might be expected to look different in meditators compared with nonmeditators. In 2005, when neuroscientist Sara Lazar at Massachusetts General Hospital in Boston and her colleagues scanned the brains of people who had spent around six hours practicing insight meditation every week for the previous ten years or so, they discovered something remarkable. Although the thickness of the cortex is known to decline with increasing age in the general population, particularly in the frontal lobe, a specific part of these veteran meditators' orbitofrontal cortex appeared to be defying this trend. This region of the brain was just as thick in forty- to fifty-year-old meditators as it was in twentysomething meditators and in control subjects in their twenties who had never meditated. Subsequent research by other neuroscientists has borne out these findings.[3-5]

The region in question is known as Brodmann area 10, and it may be an important node in a metacognitive network that allows us to observe the emotional currents of our minds and decide whether we want to swim against them or allow ourselves to be swept along.[6] Brodmann area 10 is the farthest outpost of the orbitofrontal cortex (see Figure 7, page 223), which as we saw in the previous chapter plays a crucial role in mentalizing ability, also known as theory of mind (the capacity to put oneself in other people's shoes and see the world from their perspective, a vital skill for social cognition and language). Recall that the overall volume of the orbitofrontal cortex correlates very closely with the maximum size of a primate's social group, and that it has expanded enormously over the course of human evolution.

So it's intriguing to discover that Area 10 has been emerging as a key player in the neuroscience of meditation in recent years. In 2014, a team of psychologists led by Kieran Fox at the University of British Columbia in Canada published a review of twenty-one brain-imaging studies involving approximately three hundred meditators. When they looked for areas that consistently correlated with meditation experience, they found that Area 10 had been singled out in research involving a wide range of meditation

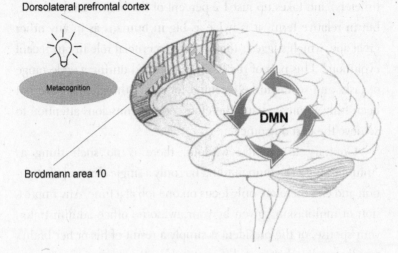

Figure 7. Metacognition. In people who practice meditation, distinctive changes have been identified in a matchbox-size region of the orbitofrontal cortex known as Brodmann area 10, which may be involved in metacognition: the ability to observe one's own thoughts and emotions with detachment. A neighboring executive control region, the dorsolateral prefrontal cortex, may work in concert with Area 10 to facilitate efficient metacognition, suppressing the self-referential thinking that is the specialty of the default mode network (DMN).

traditions; this led Fox and his colleagues to conclude that it must be fundamental to the practice.[7] Among people who meditate, studies found not only increased cortical thickness here but also greater density of gray matter (nerve cell bodies) and enhanced structural integrity of white matter (long, connective nerve fibers or "axons").

Fox and his colleagues speculate that these changes may enhance a person's ability to observe their thoughts and emotions with cool detachment, but the exact function of Area 10 remains something of a mystery. It's the size of a matchbox (around 14 cubic centimeters) and takes up just 1.2 percent of the brain's total volume, but in relative terms it is twice as big in humans as in any other great ape, which suggests that it played a crucial role in our recent evolution.[8] This part of the brain is activated during a wide range of tasks and seems to be essential for doing things that have not been fully automated and therefore require conscious attention to achieve the best outcome.[9]

Contrary to popular wisdom, there is no such thing as "multitasking": the human mind has only a single channel of attention and can therefore only focus on one job at a time. Any impression of multitasking given by your awesome office administrator, your spouse, or the president is simply a result of his or her brain's rapidly toggling between different tasks without messing up.

How does this work? For now nothing has been proven, but according to the "gateway hypothesis" proposed by neuroscientist Paul Burgess and his colleagues at University College London in the UK, Area 10 facilitates multitasking by acting like the points on a railway track, switching between two alternative kinds of cognition: thoughts that draw on memories, as happens during social cognition, daydreaming, rumination, and creative thought; and those that are focused on sensory stimuli coming from the environment and within the body.[10] These alternative cognitive routes can be conceived as a self-referential "narrative" mode and a "being" mode, respectively. As we have seen, the latter also involves the anterior cingulate cortex, insula, and somatosensory cortex, while the former involves the default mode network. When it is working well, the process of switching between the two modes of thought occurs rapidly and efficiently, for example to allow the retrieval of memories during a conversation while listening to what the other person is saying.

In the light of this hypothesis, it makes sense that all kinds of meditation that involve consciously focusing attention—whether on a mantra, a movement, an object, the breath, or a bodily sensation—and gently bringing it back whenever the mind wanders have measurable effects on Area 10. What may be happening is that this mental exercise enhances the efficiency of the switching mechanism. Picture it as an old-fashioned railway junction with a big lever controlling a set of points that determine whether trains go down one track or another. If the trains represent packages of information and the tracks are the alternative pathways through the brain that they can take, meditation would be analogous to the signalman over time developing stronger biceps and triceps with repeated lever pulls, allowing him to control the points with increasing ease. As a result, switching the stream of thought between the narrative and being modes becomes more and more flexible, fast, and efficient.

Burgess and his colleagues speculate that if a person's Area 10 "switchpoint" were faulty, this would make it difficult for them to differentiate between thoughts and sensory experiences—which is what happens during auditory and visual hallucinations. To extend the railway metaphor, the points could get switched inappropriately or become stuck in one position, sending trains down the wrong track. Some evidence has in fact been found of atrophy in Area 10 and reduced functional connectivity between it and the default mode network in the brains of people with schizophrenia.[11]

Researchers have only just begun to explore potential connections like this between Area 10 and mental health problems, but we can be certain this matchbox-size outpost of the prefrontal cortex does not act alone. The brain operates as a system of interlocking networks, each comprising several regions. As we saw in chapters 6 and 9, for example, overactivity in the default mode network, with its widely distributed components that swing into action whenever we are not focused on performing a task, has been implicated in a range of mental illnesses.

Another important player in metacognition is the dorsolateral prefrontal cortex, an executive control region that fulfills a vital role in working memory, planning, and decision making. Interestingly, it has strong functional and anatomical connections with Area 10.[12] In fact, they are neighbors, the dorsolateral prefrontal cortex occupying a region a little higher up on the outer surface of each hemisphere (see Figure 7, page 223). Like Area 10, it has massively expanded during the course of hominin evolution and, tellingly, has also been implicated in mental illness. Brain-imaging research suggests that in people in the throes of depression—and even those in remission who are at high risk of relapse—activity in the dorsolateral prefrontal cortex is sluggish compared with control subjects, whereas the ventromedial prefrontal cortex—part of the default mode network—is overactive.[13-15] It is as if the brain has somehow got "stuck" in the narrative, self-referential mode of thinking. By contrast, skillful metacognition—which involves switching efficiently and appropriately between the narrative and being modes—seems to protect people who have experienced depression from suffering a relapse.[16] Mindfulness training, in addition to beefing up Area 10, has been shown to boost activity in the dorsolateral prefrontal cortex, which may help counter the narrative tales spun by the default mode network.[17] These changes in the brain may be what underlies the success of mindfulness-based cognitive therapy (MBCT), which was explicitly designed to prevent relapse by reducing "cognitive reactivity," the tendency for small disturbances in mood to get drawn out by rumination in people who are prone to depression.

Once again, it is worth remembering that the symptoms of mental illness are more widespread than is commonly supposed and exist on a continuum of severity throughout the population.[18, 19] It may therefore be overly simplistic to talk about the "mentally well" as opposed to the "mentally ill" as if people could be definitively assigned to one group or the other. So when neuroscientists discover that the brains of people who have been diagnosed with these conditions have particular structural and functional characteristics, it would probably be more helpful to think of these as extremes on a spectrum of variation that extends throughout the general population. By the same token, when researchers find evidence that meditation affects regions associated with mental illness, the potential benefits of practice will certainly not be restricted to "patients." They will apply to anyone struggling with the everyday foibles of the human mind: its tendency to wander and get lost in emotions, its selfish drives and endless capacity for taking things personally.

We can glean fascinating clues about the changes that were wrought in Siddhārtha's brain by dedicated meditation practice in the years after he embarked on his spiritual quest—and their "wonderful and marvelous" effects on his mind—by scanning the brains of experienced, older meditators like those described at the start of this chapter. Area 10 and the dorsolateral prefrontal cortex are among the regions emerging as key players in both mindfulness and mental health, but there is another area that's been shown to be transformed by years of meditative practice. Remarkably, this easily overlooked part of the brain may be involved in the creation of conscious awareness itself.

The insula, tucked inside a fissure dividing the frontal and parietal lobes from the temporal lobe (see Figure 2, page 87), was already known to be involved in monitoring the internal state of the body, or "interoception." But, according to Bud Craig, a neuroanatomist at the Barrow Neurological Institute in Phoenix, Arizona, it is also the seat of conscious awareness, or as he puts it, "the feeling of being alive." His theory is that the insula creates a series of "snapshots" of our internal and external environment, which it progressively integrates with inputs from other parts of the brain responsible for emotion, motivation, decision making, social cognition, and regulating the physiological state of the body.[20] This mapping process is completed in the anterior or frontmost part of the insula, where information is incorporated from the anterior cingulate cortex (ACC), which detects errors while we're performing a task, monitors conflict between competing tasks, and handles volition or "directed effort." In essence, Craig argues that by integrating all these different sources of information, the insula creates the feeling of being alive. In support of his radical theory, Craig cites a wide range of evidence, including connectivity between the insula and relevant parts of the brain and the observation that the ACC and anterior insula are jointly activated when we experience every conceivable kind of emotion, from maternal and romantic love, empathy, happiness, and sexual arousal to anger, fear, sadness, disgust, and aversion. The insula has even been implicated in our enjoyment of music, the religious experience of "union with God," and the hallucinogenic state induced by ingesting the psychoactive brew ayahuasca.

Meditation techniques such as attending to the breath and the body scan aim to sharpen awareness of bodily sensations, so it makes sense that these practices lead to structural changes in the insula. As we saw in chapter 4, "The Second Dart," the insula becomes active when experienced meditators reduce the unpleasantness of pain by tuning in to the sensation with an attitude of acceptance and curiosity. The studies reviewed by Fox and his colleagues suggest that even a short meditation course can lead to greater thickness of the cortex and gray matter density in this region. And these changes are accompanied by greater cortical thickness and white matter integrity in its partner in conscious awareness, the ACC.[21] Further evidence is provided by findings from fMRI studies showing that the insula and ACC are more active in meditators than in controls, not only when they are practicing mindfulness but also during a resting state.[22] So it appears that training this circuitry is fundamental to the benefits wrought by meditation. During breath-focused meditation, for example, the anterior insula becomes active when the sensations of breathing in and out are made the subject of conscious awareness, whereas the ACC notices when your attention has wandered and brings it back on the orders of the executive control center, the dorsolateral prefrontal cortex.[23] In fact, neuroscientists consider the ACC and insula to be the core nodes of the brain's "salience network," responsible for directing conscious attention to important stimuli. Repeated practice may steadily strengthen the network so that present-moment awareness can be achieved with greater ease until it becomes second nature.

The ability to focus attention on incoming sensory and physiological signals from the body—interoception or "body awareness"—may be fundamental to our sense of well-being. It acts as a reality check, performed by the insula. By contrast, dissociating from these bodily sensations has been linked to a wide range of mental illnesses, including anxiety, depression, addiction, eating disorders, chronic pain, and PTSD. One of the most surprising findings to emerge from neuroscience in recent years is that rather than responding in real time to the vast amount of incoming sensory data, the brain tries to keep one step ahead by constantly predicting what will happen next. It simulates a model of the immediate future based on what has just happened. When its predictions turn out to be wrong—for example, we're feeling just fine then suddenly experience a stab of anxiety about a romantic date—this mismatch creates an unpleasant sense of dissatisfaction that we can either try to resolve by ruminating and then doing something to alleviate the anxiety (canceling the date, perhaps) or by updating the brain's model of reality (investigating and accepting the new sensation). These alternative strategies employ the "narrative" and "being" modes of thought I described earlier in this chapter. Of course, both strategies have their place according to the situation, but an overreliance on avoidance behavior rather than acceptance stores up problems for the future because there are many things in life that cannot be changed and therefore need to be faced. Mindfulness through interoception is all about accepting the way things are. When we are mindful, the insula continually updates its representation of our internal world to improve its accuracy by reducing discrepancies between expectation and reality. As we've seen in previous chapters, this reality check—the focusing of dispassionate attention on unpleasant sensations such as pain or anxiety—loosens the hold that they have over us. So the structural changes in the brains of highly experienced meditators of Siddhārtha's caliber, in particular in their insula and ACC, may be responsible for the imperturbable calm and acceptance that is the ultimate goal of contemplative practice, sometimes described as enlightenment or nirvana.

For years, perhaps because of its hiding place deep inside a fold of each hemisphere, the insula was dismissed as an archaic component of the autonomic nervous system that simply processed data about the body's visceral state. Craig's theory that it holds the keys to conscious awareness has brought it blinking into the light, and it seems the more we learn about the insula the more remarkable it seems. Evidence is accumulating that it may be responsible for profoundly spiritual or religious experiences. It has recently been implicated in the feelings of blissful, heightened consciousness that precede seizures in some people with focal epilepsy, for example. These ecstatic "auras" are remarkably similar to the experiences of religious ecstasy described by contemplatives of all faiths down the ages. Writing in the journal *Epilepsy & Behavior* in 2009, Craig and Fabienne Picard, a neurologist at the University Hospital in Geneva, Switzerland, reported the cases of five patients, including a thirty-seven-year-old electronics factory worker who had been experiencing seizures for the past seventeen years.[24] Describing the auras that preceded a seizure, he told them: "It is a well-being inside, a sensation of velvet, as if I were sheltered from anything negative. I feel light inside, but far from being empty. I feel really present." He remained perfectly in control of his thoughts and aware of his surroundings throughout. "I feel a stronger consciousness of the body and the mind, but I do not forget what is around me," he said. In the notes the patient made for his doctors about the auras, he was able to write more poetically about the experience, which clearly affected him profoundly: "My inner body arises from an unalterable bliss. I escape into the time space of my body. It is a moment of fullness in the loophole of time, a return to myself. It is an unconditional, privileged moment of inhaled sensations. My body and my head may interact differently to what every human knows."

In a similar vein, a sixty-four-year-old woman described the ecstatic aura that preceded her seizures as an "immense joy that fills me . . . a feeling of total presence, an absolute integration of myself, a feeling of unbelievable harmony of my whole body and myself with life, with the world, with the 'All.' I feel very, very, very present at that time. . . . Entirely wrapped up in the bliss, I am in a radiant sphere without any notion of time or space. My relatives tell me that it lasts two to three minutes, but for me these moments are without beginning and without end."

Picard and Craig suspected that these ecstatic experiences were caused by overactivation of the anterior insula, because nuclear imaging (using radioactive tracers) during the patients' seizures indicated unusually heightened activity there. Direct evidence of a link was harder to come by, however. That changed in 2013 when a twenty-three-year-old woman with focal epilepsy, whose seizures always began with ecstatic auras, was undergoing exploratory surgery. Her condition, which had seriously disrupted her life since she was fifteen, was not responding to drugs; the plan was to locate the focus of her seizures using electrodes placed at various locations in her brain and ask her how she felt when the current was switched on. Nervous tissue would then be carefully removed and, hopefully, the seizures would stop. Craig and Picard report that, perhaps unsurprisingly, the patient was in a foul mood during the procedure, but the moment an electrode placed in her anterior insula was activated, she perked up and said cheerily: "I feel really well with a very pleasant funny sensation of floating and a sweet shiver in my arms." She said this was exactly how she felt just before a seizure.[25]

The feelings experienced by patients during ecstatic seizures have a lot in common with the flashes of blissful insight reported by some people who meditate regularly—both believers and non-believers. The anonymous author of *The Cloud of Unknowing* described them as "the most wonderful sensations of sweetness and pleasure"—though he cautioned that their appearance could not and should not be counted upon: "If they come, welcome them; but do not rely too much on them for fear of weakness, for it will take a great deal of your strength to sustain such feelings and weepings for any length of time. Perhaps, too, you may be moved to love God for their sake, something you will recognise by the fact that you are excessively discontent when they are missing. If so, your love is not yet either pure or perfect . . ."[26] In other words, the experience was merely a pleasant side effect of the more serious business of getting closer to one's god. Siddhārtha, recalling what happened on the night of his enlightenment, described the first two of four progressively deepening stages of meditation (*jhanas*) as "rapture and pleasure . . . but the pleasant feeling that arose in this way did not invade my mind or remain." In the third jhana, he said he experienced "a pleasant abiding," but by the fourth jhana he had gone beyond pleasure and pain, experiencing "purity of equanimity and mindfulness."[27]

Euphoria is not a goal of the Noble Eightfold Path, though it may arise unbidden along the way. Consumed in a suitably conducive environment, drugs such as cannabis and ecstasy can produce similar sensations, and neuroimaging studies indicate that the anterior insula is involved in these experiences, too.[28] What ecstatic seizures and the euphoria of meditation or taking recreational drugs seem to have in common is a feeling of intense well-being and connection—to everything and everybody—as a result of immersion in the present moment and dissolving of any sense of having a separate self. Even for those who don't profess any religious faith, the sensations can be so unexpected and powerful that they create the impression that life makes perfect sense for the first time.

In his book *Waking Up*, atheist writer Sam Harris describes his experience of taking ecstasy at the age of twenty—not at a rave but while calmly conversing with a friend sitting next to him on a sofa.[29] Harris says he experienced a "moral and emotional clarity unlike any I had ever known" and felt sane for the first time in his life. This wonderful sense of well-being seemed to stem from a total loss of self-consciousness. He was no longer anxious, self-critical, in competition with his friend, worried about making a fool of himself. His sense of selfhood and the otherness of his friend had vanished, and with it any feeling of being judged. "I was no longer watching myself through another person's eyes." The experience was so profound, reflecting on it later Harris would realize he had been wrong to dismiss historical religious figures such as Jesus, Lao Tzu, and the Buddha as "epileptics, schizophrenics, or frauds." His views about the world's organized religions were as acerbic as ever, but now he had tasted for himself something of the psychological insight these visionaries must have experienced.

All this from consuming a chemical that excites a structure deep within the brain. What do such experiences and those of epileptics and contemplatives tell us about the mind? We don't know all the details, by any means, but the insula/ACC, Brodmann area 10, and the dorsolateral prefrontal cortex appear to switch the mind flexibly between the "being" and the "narrative" modes of thought. To return to the railway analogy, the tracks diverge at some point, with one line taking us on what might be described as the scenic route past the wonders of pure, conscious sensation: the sights and sounds, feelings—even bliss—conjured by the anterior insula. The other line regales us with stories about ourselves—the tales of mental time travel, social intuition, and imagination that are the musings of the default mode network. The marvel is that we can learn to control at will the signals that determine which track we take at any particular moment. By honing our powers of attention and emotion regulation through mindfulness practice, we can, if we wish, restrict the time we spend in the self-focused, narrative mode of thinking that can lead to anxiety and depression. We can choose to take the scenic route, favoring a more experiential mode of being in which we are not held captive by our thoughts but rather treat them as transient mental events. Regardless of where or how our minds do this, there is an open space of conscious awareness available to us where the clarity of metacognition becomes possible.

As we saw in the previous chapter, when our brains evolved the hardware needed to cope with life in very large social groups, there were some serious downsides. The practice of mindfulness helps us tune out the self-referential, narrative mode of thinking and tune in to awareness. Describing his own experiences with mindfulness, Jon Kabat-Zinn described it as a joyful return to a more innocent way of being: "When you discover meditation you get this sense of, 'I remember this! I felt like this when I was a child! I felt completely integrated, I felt completely whole.' Then you discover you can live that way as an adult. That's a profoundly liberating and illuminating thing to do. Everything can be held in awareness in that way. So we're not talking about some special, magical state, we're talking about awareness, pure awareness . . . that's what we are cultivating with mindfulness."

The alternative, narrative mode of thought was recently found to have a lot in common with dreaming. A meta-analysis of previous brain-imaging research revealed that a large portion of the default mode network is activated during dreams, whereas executive control areas such as the dorsolateral prefrontal cortex and ACC are relatively quiescent.[30] During our dreams we drift far from reality into bizarre realms where we can meet and converse with deceased relatives or fly down endless flights of stairs, yet we hardly ever question the experience or become aware that we are dreaming. There is a complete lack of metacognitive awareness. We are not quite so gullible while daydreaming, though we are often unaware that we have lost sight of the task at hand and that our mental simulations may fail to accord with reality. We are dreaming with our eyes open.

Of course dreams can be a rich source of ideas and solutions, grand plans, and fabulous fantasies. Dmitri Mendeleev famously "dreamed up" his periodic table of elements while asleep.[31] Biologist Otto Loewi was dreaming when he had an idea for a groundbreaking experiment that would reveal the chemical basis of nerve transmission (and scoop him a Nobel Prize).[32] But, as in a dream, when our minds wander we often fail to check our thoughts, feelings, and emotions against reality. We ruminate and worry about things we can't change. We forget to live in the present. As the Vietnamese monk Thich Nhat Hanh writes in *The Heart of the Buddha's Teaching*:[33] "If you don't have mindfulness your whole life will be like a dream."

Guided Meditation: Body Scan

Many of us have a strong tendency to live in our heads, losing touch with the body and its ways, treating it almost like a foreign country. We dismiss the body as little more than a troublesome appendage of the mind, when in fact there is continual two-way regulatory traffic between the two. This interchange only works well if the mind is provided with up-to-date, unbiased feedback from the body. Present-centered, nonjudgmental body awareness promotes healthy emotional and physiological regulation by supplying timely, objective information to the brain. Conversely, if the mind tries to call all the shots, this can create a positive feedback loop between emotions and our mental reactions to them, stoking anxiety and suffering, craving, and anger. We need to learn how to "let go."

The body scan is simplicity itself. Take off your shoes and lie flat on your back on the floor. Feel free to use a rug or mat to make yourself comfortable. Stretch out your legs, uncross your ankles, and lay your arms by your sides with the palms of your hands upward. You can close your eyes, or leave them open if you think you might fall asleep. Allow your breathing to slow. As your abdomen rises with each inhalation, notice the points of contact between your body and the floor, and as you exhale, feel your body sinking more deeply into the mat or rug, perfectly supported and at rest. Imagine the air flowing in and ebbing away throughout your body as you breathe in and out, like waves on a beach. Don't try to change the breath in any way, just allow it to be the way it is.

On an out-breath, allow your attention to flow all the way down through your body to your toes and let it rest there for a little while. Investigate what you can feel there as you continue to breathe in and out: any sensation of warmth or heat, tingling or pulsing, perhaps the feel of the fabric of socks or stockings on skin. Try focusing on the big toes, their nails, skin, then see if you can single out each of the other toes in turn and the spaces between them. If you don't discover any particular sensation, that's perfectly okay. You are simply noticing and accepting whatever you find. Don't make any attempt to change it. Repeat the procedure as you shift your attention to the balls of the feet, the soles, the heels where they are in contact with the floor, the tops of the feet, the ankles. Spend as long as you like on each.

Now move up through the body following exactly the same procedure of investigating whatever sensation is present. Move from one location to another, spending as little or as much time as you like: your calves, your knees, the backs of your knees, thighs, pelvis, abdomen. There's no need to mentally elaborate on any sensation of discomfort or tightness, trying to explain or control it. Just explore it, accept it, and then let it go. Move up to your chest, your shoulders, down to your hands, holding in awareness each finger in turn. Move up in stages to your head: your mouth, tongue, nose, jaw, ears, and scalp. Finally, take in the body as a whole and its points of contact with the floor, its location within the room. Take a moment to enjoy this sense of calm and interconnectedness with everything.

The body scan is a very relaxing meditation, so if you do fall asleep, there's no need to feel guilty. Perhaps next time you can try with your eyes open or sitting upright in a chair or on a cushion on the floor.

As in any meditation, you take the body scan with you

throughout the day. At moments of stress or anxiety—or if there's nothing in particular going on to hold your attention—it can be soothing to bring your attention into the body, particularly the hands and feet, allowing it to rest there for a while.

CHAPTER ELEVEN

MIND MIRRORS

Make your mind a fortress and conquer Mara with the weapon of wisdom. Guard your conquest always.
—The Dhammapada (translated by
Eknath Easwaran), verse 40

"I was hanging out with this bunch of dopeheads on the beach," the middle-aged man tells his audience. After graduating, his twenty-one-year-old self could have walked straight into a magnificently paid job for life: his godfather was a partner at De Beers. But diamonds were not exactly Jeremy's thing. Instead, here he was seven thousand miles from home on Parangtritis Beach in Java with a group of other white, middle-class Westerners who had dropped out. "I had left England on a one-way journey to the East to try and discover some kind of spiritual direction," he says. Things were not turning out quite as he had hoped. "My head was spinning with all these chemically assisted confusions and the idea that here I am on my spiritual quest, trying to pursue this mystical lifestyle, and yet there was this mess of thoughts and a kind of jangled, ragged anxiety."

It was getting toward sunset one evening when he set out by himself along the sand to visit a shaman named Dalang whom Jeremy and the other beach bums had befriended. The shaman lived in a cave, about fifteen feet above the beach, that the sea had hollowed out of a rocky outcrop thousands of years ago. Perhaps Dalang would have some words of wisdom to soothe his frazzled mind? Approaching the cave he saw a man sitting near the entrance in the full lotus posture. But it wasn't Dalang. "The sun was setting so there was this golden light. He looked basically like a surfer dude—blond hair, beard, bare-chested, sitting there in a pair of shorts, absolutely poised—but as I got closer the contrast between this perfectly still, serene figure and the rattling mess in my mind, it got stronger and stronger." As Jeremy climbed up to the cave, he says, "there was this extraordinary feeling of, 'That's what you need to do. That's where you need to be.' "

At that time he'd never seen a Buddha image, he had no connection with Buddhism whatsoever and hadn't met anyone who meditated. "I didn't want to disturb this serene, holy being. So I just sat down, leaning against the rock wall of the cliff. And after a while he opened his eyes and they were exactly the same color as his shorts, this bright, turquoise blue, and he looked at me, smiled, didn't say anything and I was like . . . so full of a thousand things I wanted to say, I wanted to ask, so small and wretched and frazzled I couldn't say anything. So I just looked at him and looked at the sunset, and after a while—I can't remember whether he got up and walked away or I did—the moment passed. I never found out who he was or where he came from. Maybe he was a deva! I don't make any claims, but that particular day, that particular presence touched something in me that said, 'Yes! That's where you need to be. You don't have to dwell in this confused, jangled mess of feelings.'

"I didn't get quite to the point of thinking that if I stopped consuming my"—he struggles to find an appropriate euphemism—"*herbal mood adjusters,* that might have an effect. I was still convinced that the more stoned you were the closer you were to reality. I hadn't quite got past that particular view, but eventually I did." The powerful image of the serene meditating figure stayed with him and within months led him to Wat Pah Nanachat, the International Forest Monastery, in northeast Thailand, where two years later, in 1979, meditation master Ajahn Chah ordained him as a Forest monk and the beach bum Jeremy Horner became Bhikkhu Amaro.[1]

Scientists have unpicked the firing mechanism of isolated nerve cells, cracked their signaling code, and are using increasingly sophisticated technology to interrogate the workings of the brain. But the operation of the three-pound organ we all carry around inside our skulls remains largely a mystery. If you didn't already know its wonderful and marvelous qualities, such as consciousness, metacognition, language, love, and creativity, you couldn't begin to predict its abilities from an examination of its individual components. The brain is so much more than the sum of its parts, not least because each of its 100 billion nerve cells is an information processing unit in its own right, interconnected with thousands of others, and each densely populated region is just a single node in many interconnected networks. No wonder neuroscientists are still puzzling over what their MRI images and EEG traces actually mean. The scale of the challenge facing them is breathtaking. It's as if upon hearing an orchestra perform a violin concerto you were to attempt an "explanation" of the music by dismantling each instrument in turn. You would end up with a tangle of catgut, scraps of varnished wood, and metal tubing, still none the wiser.

Like Jeremy Horner, Britta Hölzel was fortunate to discover meditation at a young age, beginning her practice at a Yoga ashram in India during a gap year before university. Unlike Horner, however, she didn't pursue a monastic career but has spent the past decade using science to investigate the effects of meditation on the brain. She now works at the Technische Universität München (Munich Technical University) in Germany. "In the meditation research community, myself included, we really do believe that it does good to your body, your physical health, and we believe that it must have beneficial effects on your brain as well," she says. Nonetheless she realizes that, like any scientist, she brings preconceptions to her work. She knows all too well that her own brain has been hardwired by evolution to see patterns where there may be none. This is a particular problem when interpreting data from brain scans. "We are trying to find something or coming with our own ideas—that's true for all neuroimaging research—but we are still at the very, very beginning of understanding. There is so much more work to be done. I'm thinking that when we look back in twenty years we'll just laugh about some of the ideas we had."

By focusing on the music of the mind rather than the instruments playing it, Buddhists have won a headstart over neuroscientists of a few millennia. Their approach has always been reassuringly analytical—almost scientific. Monks and nuns, if you will, are like cognitive psychologists sitting in the seclusion of their cells probing the relationship between their thoughts, feelings, behavior, and well-being. "We are not statistical scientists," says Ajahn Amaro, who before embarking on his spiritual quest in Asia dipped a toe in mainstream science, earning a degree in psychology and physiology from the University of London in the 1970s. "We don't run labs and such, but our way of practicing and teaching meditation is very experientially based—it's what actually helps people. We are working with the mind to understand it better and train it to be more malleable and function in a more balanced way." Novice monks, like students of science, are encouraged to be skeptical of received wisdom and think for themselves. "It's a fundamental principle," says Amaro. "In the Kalama Sutta[2] the Buddha says, don't believe what I say, don't believe what is 'logical' or handed down to you by tradition. But whatever is beneficial to you and to others, that is praised by the wise—take it and use it. And what is harmful to you, obstructive, and makes difficulty, well, then leave it aside. So you are the arbiter of what is true and good by your own experience and you learn your own lessons."

He values the insights about meditation provided by modern psychology and neuroscience, not necessarily for their own sake but for the pragmatic reason that most non-Buddhists are more likely to be swayed by the objective research of scientists than by the subjective reports of monastics. At MIT in 2003, at the historic first public dialogue between researchers and Buddhist contemplatives—among them the Dalai Lama—he coined the memorable expression "the Great God of Data" to describe the hold that science now has over much of humanity: the belief that something isn't worth doing until its high priests have sanctified the practice in an MRI scanner or a randomized controlled trial.[3] Despite his reservations, he is happy to attend scientific conferences and contribute to dialogues if it will promote research into mindfulness, compassion and loving-kindness, and a wider dissemination of the Dhamma: the universal truths about human existence he believes the Buddha discovered.

Mindfulness research conferences, particularly those attended by Jetsun Jamphel Ngawang Lobsang Yeshe Tenzin Gyatso, aka the fourteenth Dalai Lama, draw worldwide press attention and a celebrity following. Hollywood actors Goldie Hawn and Richard Gere were in the 1,200-strong audience at that first meeting of minds in Cambridge, Massachusetts, more than a decade ago. But this wasn't the first time the Tibetan spiritual leader and scientists had put their heads together: they had been having informal, private discussions for years, culminating in the establishment of the Massachusetts-based Mind and Life Institute in 1987 by neuroscientist Francisco Varela and businessman Adam Engle, which put the conversation on a more formal footing and set the stage for public dialogues over the ensuing three decades. Along with around 1,700 scientists and clinicians from 134 countries, I attended the institute's recent symposium for contemplative studies in October 2014 at the Marriott Hotel in Boston. On the second day of the four-day meeting, the Dalai Lama was scheduled to share a platform with neuroscientist Richard Davidson and psychologist Amishi Jha. Overnight, an entire floor of the hotel had been locked down and a security barrier erected across the wide corridor leading to a ballroom the size of a football pitch where the event was due to take place in the morning. At 7:00 A.M., unsmiling security guards in dark suits and black ties began ushering delegates through airport-style metal detectors, moving in to conduct more intimate, head-to-foot scans with handheld devices whenever someone's house keys or loose change set off the alarms. An information card delivered to every delegate's room at the Marriott recommended that we arrive at least an hour before the event was scheduled to start, and instructed us to carry a government-issued photo ID and refrain from bringing any of a long list of items, including rucksacks, umbrellas, liquids, and "weapons or tools of any kind." At the close of a meeting the evening before, the chief organizer of the symposium reminded the assembled scientists, clinicians, monks, and nuns that they would have to leave their guns behind in their rooms if they wanted to attend the dialogue. It was a joke, of course, but the threat of an assassination attempt was real enough.

Inside the security cordon, the atmosphere was an odd mixture of calm and excited anticipation as delegates lingered around the entrances to the ballroom absorbed in animated discussions with colleagues and new friends. They knew they had a long wait ahead of them. Several others found quiet corners to practice their yoga or sit cross-legged meditating on the carpet in the corridor. Needless to say, thoughts of violence were far from our minds by the time the Dalai Lama took to the stage shortly after 9:00 A.M. and was introduced by the president of the Mind and Life Institute, Arthur Zajonc, to rapturous applause. One of the stated aims of the dialogue was to discuss what guiding principles should govern the application of mindfulness in settings such as business and the military. For several years, Mind and Life Fellow Amishi Jha, sharing the platform with Davidson and the Dalai Lama, had been investigating the effectiveness of a mindfulness training program designed to boost the resilience of US Marines, with grants from the US Department of Defense. But tact and diplomacy prevented any strong opinions from emerging. This was a dialogue, not a debate. Asked later in the discussion about how his interest in science originally developed, however, Gyatso said he'd been fascinated by technology since his childhood in the vast Potala Palace in Lhasa. Modern technology had yet to reach Tibet at this time, but he recalled being given a clockwork toy British soldier with a gun that he played with happily for a few days before taking it apart to see how it worked. As a nineteen-year-old touring China on a state visit (this was after the Chinese annexation of his country but before diplomatic relations had broken down completely), he was captivated by the giant hydroelectric dams and metal smelting works his hosts were so keen to show him.

Over the past fifty years, Gyatso has embraced science more closely than any other religious leader and sought to incorporate its insights into the worldview of Tibetan Buddhism. In the Gelug tradition with which he is most closely associated, he has made science and mathematics compulsory in primary school education. Science has also become part of the Tibetan monastic university curriculum. His Holiness is on record as saying that he would abandon any tenet of Buddhist teaching if it was revealed to be incompatible with scientific evidence. In his book *The Universe in a Single Atom*,[4] he wrote: "Buddhism must accept the facts— whether found by science or found by contemplative insights. If, when we investigate something, we find there is reason and proof for it, we must acknowledge that as reality—even if it is in contradiction with a literal scriptural explanation that has held sway for many centuries or with a deeply held opinion or view." The passage from the Kalama Sutta that Ajahn Amaro quoted to me says as much. At the Boston mindfulness symposium, Gyatso revealed he had long since abandoned traditional Tibetan Buddhist cosmology after learning about the extraordinary discoveries of modern astronomy. It must be noted here, however, that there are some tenets of Buddhism, such as the belief that all living beings are subject to a cycle of birth, death, and rebirth, that cannot be disproved by science. I'll say more about rebirth in the final chapter, but suffice it to say here that this is one place where scientists and religious believers must politely part company. Just as no scientist will ever prove conclusively that, say, the Flying Spaghetti Monster does not exist,[5] nobody can disprove the existence of heaven, hell, gods, angels, past lives, or future lives. They can only infer that they are not very likely.

Despite these "no go" areas, there remain numerous avenues for collaboration, though one sometimes gets the impression that monastics, including the many Tibetan monks who have volunteered for research projects, are only allowing their brains to be picked and scanned in a spirit of slightly bemused compassion. With his characteristic bighearted guffaw, the Dalai Lama teased scientists gathered in the Marriott's ballroom: "I'm a Buddhist monk, I don't need science!" Explaining why he was so glad scientists had begun to investigate the benefits of meditation, however, he conceded: "When I talk about Buddhist literature, people may not listen." Echoing Amaro's sentiments, he said the public was more likely to pay attention to scientists because they were the ones with hard evidence at their disposal. He was pleased to see their work beginning to confirm what Buddhists already knew—that a well-trained mind is essential for both physical and emotional well-being. "We are not talking about heaven, God, or nirvana here," he said, "but how to build a happier society."

So what exactly have scientists proved, or is it too early to draw any firm conclusions? The best clinical evidence to date, which I described in chapter 6, supports the efficacy of mindfulness-based cognitive therapy (MBCT) for preventing relapse in people with recurrent depression. At the time of writing, the study published in *The Lancet* in 2015 is the most definitive investigation yet conducted into mindfulness: the randomized controlled trial involving 424 patients that compared MBCT with antidepressants concluded that the program was just as effective as drugs at preventing relapse over a follow-up period of two years—and significantly *better* among those who had experienced the most adversity during their childhood.[6] Investigations into other potential clinical applications are not so far advanced. To date, trials in people with chronic insomnia have produced mixed results, for example, though the most recent research has been promising, suggesting that six months after taking a specially tailored mindfulness course, 80 percent of patients were sleeping better, compared with around 40 percent in an active control group.[7] Similarly, studies of the efficacy of mindfulness-based therapies for people with bipolar disorder or psychosis are still at a preliminary stage.[8, 9]

Neuroscientist Britta Hölzel has been closely involved in developing an instructor-led mindfulness program for bipolar disorder,[10] though major clinical trials on the scale of those for preventing relapse in major depression have yet to be conducted. Critics have concerns that in conditions like these, meditation might actually trigger psychosis and other unpleasant experiences. I asked Hölzel whether she thought the fears were justified. "I have never seen any major problems like that in our classes," she said. "People can have increased anxiety with the practice, but it's something that we make the topic of the practice and work around. For example, rather than concentrating attention on the breath, we suggest they concentrate on the body, on the feet and legs." In a sense, learning to deal with unsettling emotions such as anxiety as they arise is the whole point of the practice, and in common with other psychotherapies, mindfulness may bring these problems out into the open.

It's a bit like extracting grit from a fresh wound: it can be uncomfortable but is essential to prevent infection and promote healing. In MBCT classes for depression, trained instructors assess every person coming into the eight-week course to establish whether they are ready and to exclude anyone at risk, and during the class they watch out for any problems and adapt their teaching accordingly, or provide individual help for anyone who is struggling. Meditation centers offering intensive retreats might not be so well prepared to address adverse reactions. Many actively discourage people with a psychiatric diagnosis from taking part, as staff may not be trained to identify problems or know how to respond. The absence of normal social interaction on silent retreats, the hunger caused by the monastic precept of not eating after midday, and the disturbed sleep patterns that can be a consequence of intensive meditation, which promotes wakefulness, may make adverse reactions more likely. Sleep deprivation, in particular, is a risk factor for psychosis.[11]

People who take online mindfulness courses get no one-to-one support, but these generally involve very short bursts of meditation of five to fifteen minutes which are unlikely to cause problems. At the time of writing, clinical psychologist Willoughby Britton at Brown Medical School in Providence, Rhode Island, is collecting reports of adverse reactions from a wide range of sources, but has yet to publish her findings.[12] In the meantime, the best advice may be to practice a little and often, and get advice from your doctor before going on a meditation retreat or taking a mindfulness course if you have been diagnosed with a psychiatric condition.

Despite recent concerns about potential downsides, evidence for the benefits of mindfulness interventions has been growing. In 2015, Dutch and American scientists published a review of evidence from 115 randomized controlled trials involving 8,683 people with a variety of conditions and concluded that mindfulness could be used as a supplementary treatment to alleviate physical and mental symptoms associated with cancer, cardiovascular disease, chronic pain, depression, and anxiety disorders.[13] None of the evidence was as strong as that reported for depression in *The Lancet*, because these were studies in which patients taking mindfulness courses were compared with others on waiting lists for the programs, or who received "treatment as usual"—which can amount to very little actual therapy. All medical interventions are boosted to a greater or lesser extent by the placebo effect: the power of belief and expectation to produce real physical and psychological improvements above and beyond any direct benefits from the treatments themselves (this is *faith healing* in the literal sense of both those words). Conversely, people assigned to a waiting list or treatment as usual may be frustrated that they are not receiving the therapy under investigation, which could actually worsen their condition.

Ideally, clinical trials should compare the therapy being studied with an "active" control group in which patients receive an equivalent intervention in terms of the time allocated and their expectation of its efficacy. For example, in newer studies, mindfulness courses are sometimes pitted against cognitive behavioral therapy (CBT), relaxation training, intensive health education, or an exercise program. In the past only a minority of mindfulness studies have met this criterion, though it is becoming more widespread. In 2014, Madhav Goyal and his colleagues at Johns Hopkins University in Baltimore, Maryland, reviewed forty-seven randomized controlled investigations of mindfulness interventions that *did* have active controls, finding evidence that they can improve symptoms of anxiety, depression, and pain.[14] They wrote that the effects on anxiety and depression, while modest, were "comparable with what would be expected from the use of an anti-depressant in a primary care population but without the associated toxicities." The evidence for reduced stress and improved quality of life was weaker: there were simply too few high-quality studies with active control groups to draw any firm conclusions. As a result of the placebo effect, trials that compare the effect of a treatment with an active control intervention tend to record smaller improve-ments than those that compare it to patients on a waiting list or receiving treatment as usual.

Designing an active control condition that is equivalent to meditation in terms of patients' expectations of efficacy is chal-lenging, not least because the public profile of mindfulness has been raised so high in recent years. One promising possibility is to compare genuine meditation with "sham meditation," in which patients are led to believe that they are going to meditate and then simply told to sit quietly with their eyes closed. In 2010, a study was published that compared the effects on psychological well-being and cardiovascular health of an hour of mindfulness training versus an hour of sham meditation training, each spread over three days.[15] None of the eighty-two undergraduate participants had any prior experience of meditation. The genuine intervention was found to be more effective than the phony one at reducing negative mood, depression, fatigue, confusion, and heart rate. Nevertheless, one could argue that sitting for twenty minutes with your eyes closed without a clue how to stop your mind from wandering all over the place will probably be a dull or even unpleasant experience that is unlikely to inspire confidence.[16]

Established, "gold standard" treatments for a particular condi-tion make the best active controls, so the results from a study investigating the efficacy of mindfulness for treating drug addiction are particularly exciting. In chapter 7, "Fire Worshippers," I described a randomized controlled trial that compared mindfulness-based relapse prevention (MBRP) with CBT and the gold-standard treatment for recovering addicts, the Alcoholics/Narcotics Anonymous 12-step program. A total of 286 people were randomly assigned to one of the three interventions. Both CBT and MBRP proved better than the 12-step program at keeping former addicts clean and sober, with CBT having the edge over mindfulness after six months. After a year, however, the people who took the mindfulness program were the least likely of the three groups to have relapsed. They reported significantly fewer days of drug use and significantly decreased heavy drinking.[17]

That's terrific news, you may be thinking, but how about benefits for people who haven't been diagnosed with a mental illness or addiction? Will practicing meditation improve their psychological well-being or enhance cognitive function? In this book I have argued that mental illnesses are extreme manifestations of traits that are our common inheritance as human beings. We are all to a greater or lesser extent prone to the effects of hostility, rumination, anxiety, paranoia, and craving. The growing evidence that mindfulness can reduce chronic pain and anxiety, prevent relapse into depression, and reduce drug craving strongly suggests that the practice has potential benefits for everyone, for example by improving emotional stability and helping us make better choices in our everyday lives. A review published in 2012 by German psychologist Peter Sedlmeier and his colleagues, which focused on 163 studies of the effects of meditation on nonclinical populations, suggests this is indeed the case.[18] It concluded that the strongest benefits were seen in terms of interpersonal relationships, levels of anxiety, neuroticism, and powers of attention. There were smaller gains in intelligence, positive emotions, and general well-being. Once again, the most stringently designed studies that compared meditation with an active control group found smaller effect sizes than studies that used waiting list controls. Nonetheless, the results look promising and the benefits appear to go beyond the effect of repeatedly evoking the "relaxation response"—the flip side of the fight-or-flight response—because, in the studies reviewed, mindfulness led to bigger effects than standard relaxation training. There seems to be more going on than simply learning how to "chill out."

So what exactly *is* going on? In addition to inducing calm, there are four other prime candidates. We have already explored three: a change in perspective on our "self," emotion regulation, and metacognition. The fourth is an improvement in our ability to focus attention and hold it steady. This is a crucial skill for understanding and relating to reality, by avoiding being distracted or fooled by the superficial appearance of things. Buddhists identify "delusion" as one of three psychological poisons that cause human suffering, the other two being craving and aversion. Everyone can agree that someone who is easily distracted is easily deluded. In fact, if we don't want to be misled, the ability to deploy attention effectively may be more important than intelligence per se, as I will explain shortly. First, take a look at this problem:

> A *bat and ball cost $1.10.*
> *The bat costs one dollar more than the ball.*
> *How much does the ball cost?*

Easy, isn't it? There may be a doubt nagging away at the back of your mind. Why did I go to the trouble of posing such a simple challenge? Hold that thought. Many readers will already have realized that the answer is not 10 cents (some may even have calculated the correct answer, 5 cents). A few seconds spent cross-checking the answer that immediately springs to mind will reveal it to be false, because if the ball costs 10 cents, then the bat must cost $1.10, which makes a total of $1.20. Don't worry if you were fooled, you are in very learned company. When students at three of America's top universities, Harvard, MIT, and Princeton, were asked to solve this problem more than 50 percent gave the intuitive answer without a second thought.[19] Calculating the correct answer requires a little basic algebra, but the point here is not whether you can do the math but whether you take the trouble to double-check the solution that pops instantly into your head. It turned out that students who were content with the intuitive answer to the bat-and-ball test and similar puzzles, without making the effort to check, were more likely to be impulsive, impatient, and hooked on instant gratification. Commenting on this finding, Nobel Prize–winning psychologist Daniel Kahneman writes in his book *Thinking, Fast and Slow*:[20] "Many people are overconfident, prone to place too much faith in their intuitions. They apparently find cognitive effort at least mildly unpleasant and avoid it as much as possible." To make rational decisions, he concludes, you must be skeptical about your intuitions—the quick-and-dirty shortcuts that often serve us well but also lay us open to all kinds of delusions. Kahneman argues that rationality therefore depends not only on intelligence but also on our taking the trouble to pay attention.

Does meditation nurture this ability to resist the siren call of intuition? In theory it can. The "focused attention" phase of mindfulness meditation involves concentrating on a particular stimulus, such as a fixed point in the environment or the sensation of breathing, to the exclusion of everything else. This not only evokes the re-laxation response but is also theorized to improve one's capacity to give sustained attention to all kinds of everyday tasks. In the "open monitoring" phase, meditators widen the field of their attention to include any thoughts and sensations that arise in the mind, but without elaborating on them or trying to change them. In time this may improve their ability to make rational choices by honing their powers of metacognition. The idea is that, in common with learn-ing a physical skill, these mental exercises change the wiring of the brain by encouraging the growth of the requisite connections and neurons, which over time makes the practice progressively more automatic and efficient. This is neuroplasticity in action, and we have seen that neuroscientists can now trace its effects in the brains of people who have been meditating for many years.

A study found that among experienced meditators who had ac-cumulated an average of 19,000 hours' practice, there was greater activity in a network of regions involved in sustained attention com-pared with novices. But among age-matched participants who had practiced for an average of 44,000 hours in the preceding decades, there was *less* activity in these regions than among novices.[21] This is the classic inverted U that neuroscientists see when they trace activity in key regions as people learn new skills: activity climbs to a peak, then tails off as performance becomes second nature. It's worth noting, however, that 44,000 hours of meditation equates to six hours' practice every day for twenty years. Fortunately, more modest but attainable improvements in attention can be achieved within months of setting out on the long road to nibbāna. One study found that three months of intensive meditation training sig-nificantly sharpened attentional focus, as measured in a standard test known as the "attentional blink task."[22] Another found evidence of improvements in sustained attention (the ability to resist internal and external distractions) after an intensive three-month retreat,[23] though the jury is still out on whether the standard eight weeks of an MBSR course is long enough to make a noticeable difference in attention skills. A study published in 2014 found no evidence of improvement after such a short time.[24]

A review of all the evidence to date by Alberto Chiesa and his colleagues at the Institute of Psychiatry in Bologna, Italy, led them to conclude that the early phases of mindfulness training, which involve sustained focus on a particular stimulus, strengthen "executive attention."[25] This is the ability to manage conflicting demands on one's limited attention resources. Later phases of training over a period of months or years, which involve open monitoring of sensations, feelings, and thoughts as they come and go without getting caught up in them, sharpen a kind of attentiveness called "alerting," in which one remains vigilant to unexpected internal and external cues without mentally elaborating on them. Chiesa and his team also found evidence that people had improved working memory after short mindfulness retreats. Working memory operates as an ongoing "buffer," allowing you to remember chunks of information such as a string of random digits for several seconds. The psychologists report that the magnitude of these improvements increased with greater meditation experience.

In addition to promoting mindful attention, other proven effects of meditation that might facilitate better decision making include improved emotion regulation (less anger or aggression, for example), reduced craving for addictive substances, and more empathy for the feelings of others.[26] This will surely be an important avenue for mindfulness research in future years, with implications that reach far beyond personal well-being into the social and economic realms. Psychologists are only just starting to probe potential links between meditation and rational decision making, but in principle, by helping people to become more alert and attentive in their everyday lives, meditation could protect them from being fooled or deluded.

When I think of the human capacity to root out delusion, I invariably recall my former boss at *New Scientist*, the late John Liebmann. As chief subeditor of the magazine, seeing through delusion was John's job, and he was frighteningly good at it. He was the scourge of writers and editors who didn't share his attention to detail, which was fueled by intellectual curiosity and gallons of bitter black coffee. Tellingly, in conversation he was often slow to respond, taking time to think through what he was about to say rather than blurting out the first thing that came into his head—which is what most of us do much of the time. Daniel Kahneman would have approved. People who didn't know John very well would complete his sentences for him—bringing a slightly pained but resigned expression to his face—but his friends and colleagues would wait patiently until the words came. We knew it would be worth the wait. As a result of his taking the time to think, John always meant what he said and said what he meant. In an argument, infuriatingly, he was usually right. John knew that the mark of a good subeditor is not intelligence or knowledge, though these help, but an instinct for when something doesn't smell right and a willingness to expend the effort to do something about it. Like Kahneman, John was fascinated by the way errors can hide in plain sight, like a typo in a headline that nobody spots until it is too late. Intrigued, he would print out an offending page and show it to each of us in turn to see whether we could find the mistake—not to assign blame but out of curiosity about the mind's blind spots. One afternoon John was doing the rounds with a printout, not a page from *New Scientist* but a mental illusion that would be featured in the magazine the following week. When he finally arrived at my desk, he knelt down (he was a very tall man) and placed the page in front of me. "How many Fs are there in this sentence?" he demanded. I looked at him skeptically. "Go on, count them!"

FINISHED FILES ARE THE
RESULT OF YEARS OF
SCIENTIFIC STUDY
COMBINED WITH THE
EXPERIENCE OF YEARS.

I scanned the sentence and responded confidently: "Three." John grinned happily. "Try again." I looked more carefully but after a few seconds I shook my head. "Still three." "There are six!" said John, clearly delighted. He explained that native English speakers tend to skip over small, common words such as *of*, barely registering them consciously. Despite my best efforts I had missed three Fs—even after being told there were more hiding in the sentence.

Nine years' learning the skills of proofreading and subediting under John's watchful eye led to what will probably forever remain my greatest contribution to British journalism. This took place in June 2008, shortly after I joined the science and technology desk of the *Guardian* newspaper in its tatty former home on Farringdon Road in East London. The team had put together a supplement about a vast machine that had been built in a circular tunnel beneath the border between France and Switzerland. It was called the Large Hadron Collider, and within a few months it would fire its first beam of protons.[27] The LHC was billed as the largest and most complex machine ever built and one of the most audacious research projects ever undertaken. Everyone now knows that after several years of bashing protons together, the collider would eventually prove the existence of the Higgs boson. Among the contributors to the special supplement were some of the biggest names in physics, including Stephen Hawking and Peter Higgs. After weeks of meticulous editing and checking, it was just hours from going "off stone"—ready for the printing presses. Some 350,000 copies would be tucked into the newspaper, loaded on trucks and trains, and distributed across the UK. I hadn't been involved up to this point, but as a new member of the team I was proud to be given the job of reading the final proofs. I remember the introductory page was headlined "Guardian Takes On the Big Bang Machine" and I noticed that across the top of this page and every other page of the supplement were printed in much smaller letters the words THE LARGE HARDON COLLIDER.

Most people read too fast to notice such things—their eyes dart through text, allowing them to consume the information in newspapers, magazines, and books at great speed. But some of us are slow readers and have been able to build careers around this apparent handicap. According to Kahneman, the source of most common delusions—mathematical, textual, social, economic, or political—is rapid, intuitive cognition combined with a failure to deploy the brain's slower, more laborious reality-checking systems. Typos are a trivial example. More seriously, his research has revealed that many of the important decisions we make as we navigate our complicated world are governed by unconscious biases. Some of these are the work of what neuroscientist David Eagleman calls the brain's "zombie subsystems": unconscious survival mechanisms that evolved to help us respond quickly in potentially advantageous or dangerous situations.[28] Your zombie subsystems will kick in when you're driving and another road user cuts you off, for example, or when you're at a party and someone accidentally shoves you, sending the contents of your glass flying. They are responsible for all kinds of fears, anxieties, and prejudices, for the ingrained belief that someone physically attractive is more trustworthy, or the intuitive conviction that a child wearing glasses is smarter than her schoolmates. The importance of sharpening our powers of attention is that it might help counter some of the nonsense that our zombie subsystems serve up, rejecting their quick-and-easy logic.

When I interviewed Zindel Segal, one of the three codevelopers of mindfulness-based cognitive therapy, he pointed out something that, like many other recent mindfulness converts, I had failed to take on board. "Mindfulness right now is presented as a sort of mental state," he said. "But at its core it's not about entering a state, it's about the ability to use a certain perspective on your experience that allows you to make better choices. Mindfulness isn't the destination, it's just that when you're mindful you're able to make decisions that are more healthful, less reactive, more adaptive, more compassionate." So the ultimate goal for someone embarking on mindfulness training must be to become more attentive to reality. This is an open-minded, nonjudgmental kind of attention, but attention per se can be nurtured in many different ways. It can be honed during early childhood simply by giving children the right kinds of games to play. Psychologists at the University of Oregon encouraged four- and six-year-olds to play specially designed video games in five daily forty-minute sessions. The games required attention and control. One involved using a joystick to move a cartoon cat around the screen to keep it within a shrinking, grassy patch of dry ground and avoid muddy areas. Remarkably, this simple training not only led to sustained improvements in the children's executive attention but also increased their scores on tests of nonverbal intelligence.[29] More than a hundred years ago, the great American psychologist William James wrote, "The faculty of voluntarily bringing back a wandering attention, over and over again, is the very root of judgment, character and will. No one is *compos sui* [master of oneself] if he have it not. An education which should improve this faculty would be *the* education *par excellence.*"[30] Paying attention to anything that might be construed as "boring" is not something that children and teenagers are known for, however, and in recent years some of the least attentive are likely to have received a diagnosis of attention deficit hyperactivity disorder. In the US, hundreds of thousands of children and adolescents are now medicated as a result.[31] Because mindfulness training explicitly hones the brain's powers of attention, it offers a potential alternative, and there is preliminary evidence from research that it might work in adolescents.[32] The sticking point may be that children now grow up expecting their senses to be continually stimulated with novelties, whereas in meditation one must learn to be satisfied with a bare minimum of stimulation.

As a result of our incessant interaction with multimedia during waking hours, many of us may be suffering from what is jokingly called "multitasking addiction disorder" (MAD). Our minds have become conditioned to expect a stream of rapidly changing stimuli and so we get bored and restless in their absence. "One purpose of learning to develop attention with uninteresting objects like the breath is to establish that quality of attention and ease with the present moment with an absolute minimum of stimuli, so that you can feel at home in yourself with very little going on," Ajahn Amaro told the audience at that first public Mind and Life dialogue between scientists and contemplatives at MIT in 2003. "Teenagers need to be playing a video game and listening to the stereo and checking their email simultaneously, and then they feel good! That takes a lot of hardware to feel at home in life." Ironically, a regime of mindful sensory austerity, by sharpening and automating executive attention, improving alerting, and extending working memory, may improve one's ability to switch efficiently between different tasks.

What unites tasks that require sustained, single-pointed attention is that they enhance our ability to silence the chatter of the brain's default mode network. This may have a more long-term effect that, if proven, would be one of the most surprising and welcome benefits of mindfulness: helping to slow age-related cognitive decline and perhaps even protect against Alzheimer's disease. The first line of evidence in support of this extraordinary possibility is that brain atrophy and deposits of beta-amyloid—the protein plaques that are characteristic of the disease—are concentrated in the default mode network, including its principal hub, the posterior cingulate cortex, and the medial temporal lobe, where structures such as the hippocampus create and store long-term memories.[33] In other words, when healthy younger people are not focused on performing an externally directed task, the parts of their brain that become active closely match the areas most vulnerable to the damage associated with Alzheimer's in the elderly. As we have seen, the brain's default or "mind-wandering" mode of operation is to draw on memories to simulate the past and the future. The second line of evidence is that in mice genetically engineered to develop amyloid plaques, these build up exclusively in areas of high nerve activity.[34] So, in theory, any pastime that holds the mind's eye steady and stops attention from wandering—whether it's a sport, solving puzzles, math, reading, studying, or the mindful awareness of everyday activities cultivated by meditation—will give the brain's default network a break and make it less likely that amyloid plaques will accumulate. Finally, there is preliminary evidence that meditation can slow or even reverse age-related brain degeneration, helping to maintain the thickness of the cortex and prevent loss of gray matter (nerve cell bodies) and white matter (nerve fibers or "axons").[35]

The idea that meditation can help prevent Alzheimer's is highly speculative, of course, and the challenges that will be involved in proving or disproving this possibility provide an object lesson in just how tricky medical science can get. We already know that people who spend a lot of time studying in their youth are less likely to get Alzheimer's in old age. Would a youth spent cultivating mindfulness have a similarly protective effect? The obvious place to start looking might be a monastery, where some of the monks and nuns will have been diligently meditating and practicing mindfulness since their teens or early twenties. You could compare them with people outside the monastery walls of similar age, educational qualifications, general health, and so on and wait to see which ones go on to develop dementia in the ensuing decades. The problem is that monastics are unusual in all sorts of ways. The Thai Forest monks, for example, are strict vegetarians and never eat after midday. Their lives are almost certainly less stressful than ours; they spend less time sleeping and are not exposed to the torrent of information and entertainment that we take for granted. In addition, they may be exposed to fewer environmental toxins than the average, and they don't smoke or drink. To make epidemiologists' lives even more difficult, monks and nuns probably had an atypical psychological profile even before they embarked on the contemplative life, as it takes a special kind of person to give up voluntarily any prospect of sex, marriage, starting a family, and acquiring all the status symbols and other paraphernalia that the rest of us consider to be so important. All these factors are potential "confounding variables" that may also affect an individual's risk of developing dementia. So you can appreciate the challenges that will face scientists trying to find out whether meditation protects people against Alzheimer's. It took four decades of intensive, costly medical research to conclude that added sugar in food and drink presents a more serious threat to cardiovascular health than dietary fat.[36] How much harder will it be to prove that regular meditation can shield us from dementia?

Even if it were proven that regular meditation protects against age-related cognitive decline and Alzheimer's, not everyone would be willing or able to set aside twenty minutes a day, year in year out, for the rest of their lives. By the same token, can taking an eight-week mindfulness course provide lasting protection against anxiety and depression if the practice isn't sustained? Popping pills—the mainstay of modern medicine—seems easy by comparison. And formal meditation is just one element of the program. The goal of mindfulness practice is to transform the way one relates to reality from moment to moment throughout the day. An easy fix it ain't. Could going on meditation retreats of the kind provided by monasteries help people maintain their practice? One person who has been doing this for years is Judith Soulsby, a mindfulness trainer at Bangor University who was involved in early research into the efficacy of MBCT. She happened to be on retreat at Amaravati while I was staying there, and Ajahn Amaro recommended I get in touch. When we spoke on the phone a few months later, I asked her whether she thinks an eight-week mindfulness course is enough to turn around somebody's life and provide lasting protection against mental health problems. "The way I see the eight-week course is that it's a way of launching people on a particular trajectory," she said. "Eight weeks is enough to give people a sense of whether they want to continue, and it gives them the skills so they can. For most it's actually very difficult to continue without any input, so in Bangor we run a follow-up group every month and quite a lot of people come to that who have been on the course. Coming back into a group seems to refresh their practice." Most mindfulness courses are devoid of any religious content, I say. Could Buddhist retreats have a place in sustaining enthusiasm over the long term? She said she had found them very beneficial herself. "Going on a retreat each year and teaching and training as I do really keeps me strongly in touch with my practice. The important thing seems to be to continue practicing, however people choose to do that."

Nevertheless, it takes steely determination to meditate regularly at home without the help of an instructor or a group of like-minded people. Audio recordings of guided meditations can help, but for many the challenges of setting aside even fifteen minutes a day to sit still and do "nothing," with only their wandering mind for company, may be too great. As it usually does, technology may come to their rescue. The future of meditation may be "mind gyms," where members will settle themselves in front of a screen first thing in the morning before work, don an electrode cap, and perform a mental workout. The technology is already being developed. Neuroscientist Judson Brewer and his colleagues at Yale University School of Medicine are working on equipment that could be used at home—or your local mind gym—to monitor activity in a target area within the brain using EEG (electroencephalography).

"You can process people's brain activity really quickly now and give them feedback on how they can increase or decrease activity in a particular brain region," Brewer told me. "We think of it as a mental mirror." During meditation the electrode cap will pick up the characteristic brainwave signature of this region and display it on a screen. By looking at your mind in the mirror, you will learn how to enter and remain in an optimum mental state to achieve the best possible effects and see straight away if your thoughts have started to wander. Unlike in a traditional gym, heaving and straining will be discouraged. "It will be about dropping into this effortless state where you're just resting in awareness and not *doing* anything," said Brewer. "You're not clenched. You are totally immersed in what's going on and enjoying the ride."

Meditation has changed little since the techniques were developed thousands of years ago in ancient India, but technologies like the one being developed by Brewer and his colleagues offer the enticing prospect of making the practices easier to learn and more effective—even entertaining. They also offer mindfulness researchers powerful new tools. Brewer and his team have used real-time feedback from fMRI scans to explore the patterns of brain activity that correspond to subjective experience during meditation.[37, 38] In their experiments, Christian and Buddhist contemplatives lay in the scanner and watched a screen displaying changes in the activity of their posterior cingulate cortex (PCC) from moment to moment as they meditated. They were later able to tell the researchers how different mental states affected the graph at particular moments. For example, when they experienced "contentment," "undistracted awareness," or "effortless doing," the graph showed falling PCC activity, whereas feelings of "discontent," "distraction," or "efforting" corresponded to increasing activity. The experienced meditators quickly learned how to control the graph at will using their minds.

This new approach, directly linking subjective experience with brain activity in real time, has become possible in recent years with improvements in MRI technology. Future studies may be able to identify the neural signature of thinking styles associated with particular problems, such as rumination in major depression or craving in people addicted to nicotine or other drugs. Clinicians could then develop games or exercises designed to help them recognize and alter these unhealthy mind states. One day we may use technologies like this routinely to optimize meditation practice and fine-tune our minds. What would the Iron Age sage Siddhārtha Gautama have made of that? What would he think of the explosion of "secular" mindfulness? After interviewing the Thai Buddhist monk Ajahn Amaro, I began to suspect he would be a little disappointed. Amaro is pleased to have seen mindfulness take root in mainstream Western medicine and culture over the past decade. But he believes secular mindfulness falls far short of the Buddha's vision for transforming human beings, because it lacks major elements of his formula for ending suffering. Only three steps in the Noble Eightfold Path—right effort, mindfulness, and concentration—relate to meditation, but secular mindfulness teachers make no mention of the other five steps, which concern wisdom and ethics. It is as if doctors were to prescribe only one drug of the cocktail of antibiotics needed to cure a patient with TB. Amaro concedes that the decision by Jon Kabat-Zinn—an experienced practitioner of Zen Buddhism—to secularize meditation when he developed MBSR in the 1980s was "skillful," allowing it to gain the widest possible acceptance among doctors and the public at that time. But he believes such interventions do patients a disservice by steering clear of any advice about how unethical behavior, such as lying or sexual misconduct, might impact their psychological well-being. "I do feel that something is lost," he says. "In therapy it is verboten to touch on any criticism of people's conduct."

The five steps on the Noble Eightfold Path missing from secular mindfulness are right view and intention (wisdom), plus right speech, right action, and right livelihood (ethical conduct). Amaro is not proposing that secular mindfulness instructors teach "the path," let alone stand in judgment over their patients. Rather, he believes they should advise them that certain kinds of behavior are detrimental to psychological well-being while others are beneficial. Unlike other faiths, Buddhism does not divide behaviors into "sins" and "virtues," but rather "skillful" and "unskillful" judged solely on their psychological effects. There is no sin in Buddhism. Writing in the journal *Mindfulness* in 2015,[39] he argued that the recommended ethical standards could be couched in terms such as: "Here are some guidelines for behavior and speech that might help you to reduce stress and live more comfortably; if you are interested you can try them out and see what their effects are." He calls his proposal "holistic mindfulness." This would remain devoid of religious content, but he nevertheless suggests as a framework for the guidelines the "five precepts" that lay followers of Buddhism set as their goal: to abstain from harming living beings, stealing, lying, sexual misconduct, and intoxication.

I asked Amaro to clarify the connection he is making between unethical behavior and psychological harm. He used the example of lying. "If you tell a lie, you have to remember that you told a lie, you have got to sustain that lie, you have to deal with the negativity that comes at you when they find out that you've lied. You have to deal with the receiving of disrespect from others when they know that you have been lying. That's a tension in your own mind. There's a stress there. You're feeling stressful and alienated and agitated, because it's a natural effect of that particular course. If you want to stop feeling that particular kind of stress then if you don't tell lies in the future you avoid creating that for yourself—it's not like saying, if you tell a lie then you're evil and you'll be punished." As a Buddhist monk, he feels no need to consult the "Great God of Data" to be confident that such connections exist between ethical behavior and well-being, but he believes science will catch up sooner or later. "My prediction is that within a couple of years ethics is going to be the new discovery [in psychiatry] and people will say, 'Wow, how amazing! What you do has an effect on how you feel about yourself! Gasp! What a revelation!' " But he concedes that ethics will remain "delicate territory" for many people.

If anyone takes up Amaro's proposal, holistic mindfulness may be difficult to test scientifically. Incorporating ethical guidance into mindfulness interventions might introduce too many variables, making it almost impossible to tease apart their effects on the symptom targeted, such as anxiety, stress, or cognitive reactivity. And some of the precepts, such as honesty or not harming other living beings, might be tricky to measure empirically. The patient would have to test their worth against their own experience, which is the essence of the Buddha's teaching in the Kalama Sutta but unlikely to appease the God of Data.

When I interviewed Kabat-Zinn a few months after my stay

at Amaravati, I asked him what he made of the abbot's idea of incorporating an ethical element into MBSR. He pointed out that while MBSR instructors don't talk about ethics in the way a monk might at a Buddhist retreat center, they nonetheless operate within the ethical framework provided by the Hippocratic Oath—summarized as "first do no harm." This is a sacred principle governing the relationship between teacher and student in mainstream mindfulness (he avoids the term *secular* for reasons I'll explain shortly). "So we prefer to embody our ethics rather than talk about it," he said. "If you only have eight weeks with people in a hospital, a clinic, a school, or whatever, the presumption of right action or ethical behavior I think is a very good place to start, and if you come across wrong action or behavior then that is held mindfully and discussed in appropriate ways. But to lecture on it and have it as a threshold to get in the door could actually alienate more people than it illuminates." He prefers to keep the ethics implicit.

His objection to the description of interventions such as MBSR and MBCT as "secular" is that it conveys the idea that the insights provided by meditation in a clinical context are qualitatively different and somehow less "sacred" than those acquired at a Buddhist retreat center or monastery. He doesn't recognize any such distinction. "We're taking this stuff that came out of the Buddhist tradition, but the Buddha wasn't a Buddhist. This was never about an -ism like Buddhism or anything like that. It was about a deeper understanding of what it means to be human. He discovered something about wakefulness and the nature of suffering and the nature of reality when you start to cultivate the mind in a certain way. That's universal."

Guided Meditation: Mindful Toast

This toast-based exercise is derived from the raisin meditation that Jon Kabat-Zinn formulated for his Mindfulness-Based Stress Reduction (MBSR) course, but before we tuck in, it's worth remembering that there is more to mindful eating than either raisins or toast—the same principles apply equally to apples, bananas, pies of various sorts, pasta dishes . . . The overarching objective of all mindfulness meditation is to learn the skill of paying attention to what is happening right here and now, to be fully present during everyday experiences and extract yourself from the dream-like stream of cogitation.

Toast a piece of bread and as it's cooking savor the distinctive aroma of white or brown, ready-sliced or crusty, seeded or unseeded. Note any happy mental associations that arise. Spread the toast with butter, marmalade, or whatever takes your fancy, listening to the rasp of the knife and appraising the texture of the toasted bread as you do so. Notice and accept any frustrations as they arise in your mind. Perhaps the butter is straight out of the refrigerator and difficult to spread? Perhaps some sticky marmalade has found its way onto your fingers?

When the toast is ready to eat, look at it as if you had never seen a piece of toast before. Notice the fine structure of the bread still visible around the edges where the spread hasn't reached, the difference in coloring and texture compared with the crust, the smoothness or runniness of the butter, the glossiness of the marmalade, its contours on the bread, its coloration and constituents. Maybe there are fine strips or chunks of orange rind, tiny air bubbles, flecks of darker color. None of this is beneath your attention.

Sniff the toast and marmalade. Take a bite. Observe how the jaws, tongue, and salivary glands immediately go to work of their own accord. Don't try to stop them or slow them down, just note the crunch of each bite and the accompanying sounds inside your head. Now that you are actually paying attention, it might be surprising how loud these sounds are. Notice the changing texture of the food in your mouth as the teeth grind it down and saliva dissolves it. Give your full attention to the sharp acidity and sweetness of the orange, the oily butteriness of the butter, the nutty toastiness of the toast.

Try following all the fully automated movements of the tongue, jaws, and lips as you chew and finally swallow. Notice the unfolding of all these behaviors as they happen, the almost unstoppable motivation to take another bite, and then another. All your impressions are valid, both the positive and negative. You may discover that you are enjoying the toast a whole lot more than if you'd just wolfed it down without thinking. Or perhaps you find the whole experience slightly disappointing. Maybe the toast is cold and chewy, the marmalade too sweet? There may be bitter, burned bits.

Accept it all with equanimity. This is simply how the toast is.

CHAPTER TWELVE

THE DEATHLESS REALM

I have gone through many rounds of birth and death,
looking in vain for the builder of this body. Heavy indeed
is birth and death again and again! But now I have seen
you, housebuilder; you shall not build this house again.
Its beams are broken; its dome is shattered: self-will is ex-
tinguished; nirvana is attained.

—The Dhammapada (translated by
Eknath Easwaran), verses 153–154

On the night of his enlightenment, Siddhārtha sat resolute and unmoving beneath the bodhi tree "as if welded with a hundred thunderbolts."[1] Seeing him there, the Angel of Death, Mara, thought to himself: "Siddhārtha the Prince wants to free himself from my dominion!" Beating his war drum, he called up a mighty army to dislodge his opponent. Gods and angels ran for their lives at the army's approach, leaving Siddhārtha to face the hordes alone.

All the homeless man had to protect him were the Ten Perfections of generosity, virtue, renunciation, discernment, persistence, endurance, truth, determination, goodwill, and equanimity.

When Death had marshaled his forces about the tree, they stretched for twelve leagues in every direction and their war cry was like the sound of an earthquake. Then the Evil One mounted his battle elephant, "Girded with Mountains," and commanded a whirlwind to blow, fit to uproot trees and tear down the peaks of mountains. When that failed to perturb Siddhārtha, he sent a great flood, then a storm of falling rocks. When these also made no impression on the ragged man sitting cross-legged under the tree, he summoned a hail of deadly weapons: swords, spears, and arrows that fell down on him from the sky. Siddhārtha remained unmoved. Then with a wave of his hand the Angel of Death dispatched a rain of ashes . . . a sandstorm . . . an avalanche of mud . . . a curtain of thick darkness . . . but as Siddhārtha sat meditating on the Ten Perfections, it was as if all these disasters of nature were transformed into showers of heavenly blossoms before they could harm him.

Then Mara, sitting astride his mountainous war elephant, approached the bodhi tree. Towering over the scrawny man sitting on his throne of grass, he commanded him: "Get up from that seat, it does not belong to you! It is meant for me!" Opening his eyes, Siddhārtha looked up at the Angel of Death and spoke for the first time: "It is not you who have perfected the ten virtues. It is not you who have renounced yourself and diligently sought after knowledge, wisdom, and the salvation of the world. This seat does not belong to you, Mara, it belongs to me. Who is witness to your charity?" Mara smirked and waved his hand to indicate the numberless ranks of his army: "There are so many!" And at that the hosts of the Evil One shouted as one: "We are his witness! We are his witness!"

Scenting victory, Mara looked down and laughed: "And you? Who is your witness?"

"You have living witnesses," replied Siddhārtha, "I have none. But let this great and solid Earth, unconscious though it is, be my witness." Then he raised his right hand from his lap and touched the ground: "Are you or are you not my witness?" And the great Earth replied with the voice of thousands of millions: "I am your witness!!" Hearing that ear-splitting sound, the war elephant fell to its knees before the Enlightened One sending Mara tumbling head over heels to the ground, and seeing their master defeated, his great army fled.

Half a cherry plum fruit. That was all that Ashoka the Great, once mighty emperor of all India, had sovereign power over in his final days on earth in the third century BCE. According to the legend, he gave the scrap of fruit to monks at a local monastery, who had it mashed up and stirred into a soup.[2,3] Knowing that he was dying, Ashoka had already donated millions of gold pieces to the many monasteries he founded across his empire during his forty-year reign and to sites of pilgrimage such as the tree at Bodh Gaya, before his ministers and heir confiscated the imperial seal to prevent him from emptying the state coffers. Undeterred, during the final hours of his life the emperor had a legal document drawn up that gifted the entire Earth to the Sangha (the monastic community), then sealed it with his teeth before breathing his last.

Historian Charles Allen in his book *Ashoka* describes the "pathetic story" of the cherry plum, but many Buddhists reading about the emperor's final days would interpret his attempts to give away every last possession for the greater good as acts of heroic generosity that surely would have earned him considerable "merit" for his next life. Buddhists have traditionally believed that every intention, volitional act, and utterance—the generous, the malicious, and the neutral—accumulates kamma during one's lifetime that will determine the realm of existence into which one is reborn. Stacking up good kamma (literally, "action"; *karma* in Sanskrit) during this life in the World of the Five Senses will help ensure you are reborn either as a human once again or into one of the higher realms. As a direct consequence of this whole setup, your wealth, social status, and health are partly a result of kamma accrued during past lives. But there is still plenty of room for the operation of free will in this life. Donations to the Sangha are said to have particularly potent effects on how things will turn out, as is your mental state in the final moments of your life. Like a busy international airport, death can fly you to many destinations: Buddhists envisage no fewer than thirty-one in total. The happy realms above the World of the Five Senses are occupied by gods, devas, and formless beings, whereas the unhappy ones below are populated by demons engaged in relentless conflict with each other, and by animals and hungry ghosts. We have already encountered this last class of unfortunate creatures, doomed to wander the earth being eaten alive by cravings they can never satisfy, in the context of addiction.

I wonder, in which plane of existence does the being who was once the Emperor Ashoka now reside? Did he earn enough merit in the third century BCE to ensure a favorable rebirth? His rise to power and subsequent military conquests were notoriously bloody and merciless, but around 265 BCE his worldview changed dramatically. He became a devout Buddhist and model ruler, promoting nonviolence, religious tolerance, and the fair treatment of prisoners across his vast empire. He had wells dug and trees planted for shade along roads. On his orders, botanical gardens and hospitals were established for the welfare of all his people. He also embarked on a program of stupa and monastery building that reached across the subcontinent, deep into modern-day Afghanistan, Pakistan, Nepal, and Bangladesh, and he dispatched Buddhist missionaries to spread the Dhamma to kingdoms as far afield as the Mediterranean. Among these missionaries were his own son and daughter, who are credited with bringing the Dhamma and a cutting of the sacred bodhi tree to Sri Lanka, where both took root. Buddhism and the descendants of the tree flourish there to this day. Ashoka's devotion to the Dhamma undoubtedly rescued the religion that Siddhārtha had established more than a century earlier from obscurity. There is an ironic footnote to the tale, however, because after Ashoka's death, his ministers are said to have "bought back" the Earth from the Sangha for a stingy four gold pieces. It has remained under state control ever since.

In the Buddhist cycle of existence, saṃsāra, beings are born, they suffer, die, and are reborn, over and over and over again. Depending on the sum of their deliberate actions, the good and the bad, they can either rise to the heights of a blissful realm or plunge down to hell. The only possible escape from this interminable game of snakes and ladders is perfect enlightenment—the liberation that is nibbāna. Ashoka must have accumulated a great deal of bad kamma in the blood-soaked early years of his reign, which will have increased his chances of being reborn into one of the dreadful lower realms, though he has had a couple of millennia since then to work off its effects. Unlike the heavens and hells envisaged by other religions, his stay in any one Buddhist realm of existence will have been temporary. Who knows, by the end of his life his many good works and meritorious acts in the final three decades of his imperial reign may have tipped the kammic balance sheet in his favor. At any rate, if the operation of kamma is as equitable as it's cracked up to be, by now he will be happily established in some lofty spiritual realm. Better still, he may have escaped the gravitational field of saṃsāra altogether and attained the deathless state of nibbāna.

Deep-rooted cultural factors in Buddhist countries such as Burma, Cambodia, Vietnam, and Thailand play a major role in how the concept of rebirth is viewed. There are revered monks and nuns who claim to remember past lives, which for them validates their belief in rebirth, but there are equally respected ones who have no such memories. To have doubts about whether rebirth even occurs is not considered heresy. In 1993, at a Western Buddhist teachers' conference in Dharamsala, India, the Dalai Lama was asked whether it is necessary to believe in rebirth to be a Buddhist. "It doesn't matter!" he replied emphatically. This took his audience by surprise. He is after all usually billed as the fourteenth reincarnation in the lineage of Dalai Lamas that can be traced back to a man living in the fifteenth century. In case there had been any misunderstanding, he said it again: "It doesn't matter!" The most important thing was to practice the essence of the Buddha's teaching—impermanence, selflessness, and compassion—he said, though he went on to assert that with increasingly refined states of meditation, one would invariably gain the insight that rebirth was real and that to escape from the cycle of suffering, one must attain nibbāna.[4] As a materialist, I can't help but speculate about how kammic information might be transmitted from one generation to the next. According to the traditional view, at the moment of death something akin to a radio signal encoding the information is transmitted from the dying being to an embryo or infant, which downloads and stores it rather like recording data on a hard disk. When the child is old enough to act volitionally, and throughout his or her adult life, the kammic data are updated and upon death the cycle repeats. Apart from subjective recollections of past lives, there is currently no evidence for such a mechanism, and you have to wonder how it might have evolved, in common with everything else we know about living creatures, through natural selection. But because nobody can prove that rebirth *does not* occur, this is another of those branches in the road where scientists and believers must part company, albeit on friendly terms.

Belief in an orderly system of rebirth or reincarnation, in Buddhism or any other religion, seems to stem from the innate human need to see justice done and from a refusal to accept that good and bad fortune might simply be the product of random events or circumstances beyond our control. When generous acts go unrewarded and malicious acts remain unpunished, justice demands that the kammic charge sheet be carried over into the next life. A logical outcome of this transmigratory judicial system is that disease and natural disasters are not purely the result of bad luck but are also punishment for misdemeanors in a previous existence. Consoling though it may be to think that tyrants, terrorists, and their ilk will suffer dreadfully for their crimes in future lives, finding themselves reborn as slugs or hungry ghosts, it is not hard to see why many Buddhists now reject the idea that natural phenomena such as leukemia and earthquakes are the result of bad kamma accumulated during past lives. Even Buddhists in traditionalist schools, including the influential Thai Forest monks Ajahn Sumedho and Amaro, dismiss any speculation about rebirth as a waste of time.[5] They believe we should instead focus on the kamma that determines our psychological well-being in *this* life. "It is important to remember that Buddhism is a religious approach based on experience. It's not a belief system," Amaro patiently reminded me when I raised the issue. "My own teachers, Ajahn Chah and Ajahn Sumedho, would emphasize the fact that the rebirth across lifetimes is just a macrocosm of the rebirths that occur day by day and moment by moment. There is nothing very mysterious to see the cause and effect that takes place as a result of your actions. Believing in past lives and future lives doesn't actually help you very much."

The Buddha taught that by cultivating insight through meditation and mindfulness we can break the endlessly repeating, kamma-driven chain of events that drives suffering. His psychological model envisages a cycle of birth and death that can be conceived either metaphorically or literally, either as the dynamo of suffering in this lifetime or as saṃsāra spanning different lives. According to this view, it is ignorance about the true nature of existence that keeps us trapped in the miserable cycle. We fail to accept or even notice that everything without exception is impermanent and therefore ulti-mately unsatisfying, and that by the same token the unchanging Self is no more than a convincing illusion—a special effect staged by the mind. The chain of cause and effect he proposed, called "dependent origination," comprises a sequence of twelve links with each link dependent upon, or conditioned by, the previous link in the series, almost like a chemical chain reaction (see Figure 8, page 282). Ignorance conditions "fabrications" (mental volitions, or kamma), which condition consciousness, which conditions "name and form" (the person), then the six senses, environmental stimuli, feeling, craving, clinging/attachment, "becoming" (striv-ing), birth (physical or psychological), and the bitter final products of the entire process, "aging and death, sorrow, lamentation, pain, distress, and despair."[6, 7] In essence the model proposes that igno-rance about the true nature of existence triggers a chain of mental

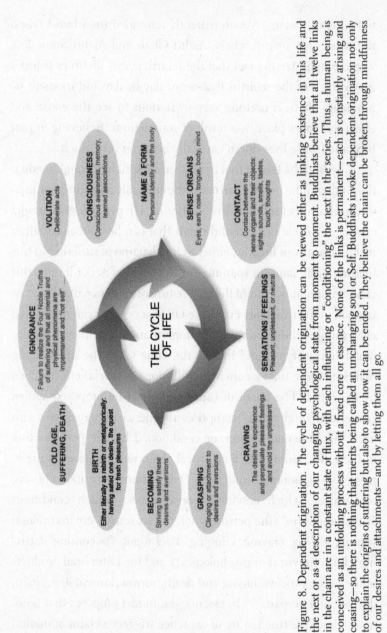

VOLITION
Deliberate acts

CONSCIOUSNESS
Conscious awareness, memory, learned associations

NAME & FORM
Personal identity and the body

SENSE ORGANS
Eyes, ears, nose, tongue, body, mind

CONTACT
Contact between the sense organs and their objects: sights, sounds, smells, tastes, touch, thoughts

IGNORANCE
Failure to realize the Four Noble Truths of suffering and that all mental and physical phenomena are impermanent and "not self"

THE CYCLE OF LIFE

SENSATIONS / FEELINGS
Pleasant, unpleasant, or neutral

OLD AGE, SUFFERING, DEATH

BIRTH
Either literally as rebirth or metaphorically: having sated one desire, the quest for fresh pleasures

BECOMING
Striving to satisfy these desires and aversions

GRASPING
Clinging or attachment to desires and aversions

CRAVING
The desire to experience and perpetuate pleasant feelings and avoid the unpleasant

Figure 8. Dependent origination. The cycle of dependent origination can be viewed either as linking existence in this life and the next or as a description of our changing psychological state from moment to moment. Buddhists believe that all twelve links in the chain are in a constant state of flux, with each influencing or "conditioning" the next in the series. Thus, a human being is conceived as an unfolding process without a fixed core or essence. None of the links is permanent—each is constantly arising and ceasing—so there is nothing that merits being called an unchanging soul or Self. Buddhists invoke dependent origination not only to explain the origins of suffering but also to show how it can be ended. They believe the chain can be broken through mindfulness of our desires and attachments—and by letting them go.

and bodily events that, via craving and attachment, culminate in our being "born into" suffering. We crave sensory pleasures that are impermanent and therefore ultimately unsatisfying, but rather than realize our mistake, we keep coming back for more of the same. We are caught fast in a trap, being metaphorically born and dying over and over again in an endless cycle of suffering.

The seeds of bad kamma that we sow in our minds are craving, lies, hatred, aggression, and so forth, and the harvest is suffering. This isn't a uniquely Buddhist insight, of course. As the Bible has it: "Whatsoever a man soweth, that shall he also reap."[8] What set the theory of dependent origination apart was its audacious attempt to explain, without recourse to divine intervention, the entirety of mental experience in a single formula: how the body, the environment, and the mind interacted to cause suffering. It asserted that x invariably followed y not because a god or the devil had decreed it but because this was the natural order of things: a physical law of existence. If you were dubious about the formula, you were perfectly free to test it against your own experience and draw your own conclusions. This was the empiricism of its day, an Eastern brand of enlightenment that preceded the first glimmerings of Western science by two thousand years. But before we get carried away and nominate Siddhārtha for a posthumous science prize, it is worth noting that he may have formulated this cyclic chain of cause and effect to reconcile his revolutionary idea about there being no such thing as an unchanging Self with preexisting religious beliefs in ancient India about rebirth—beliefs that today's scientists will find much harder to swallow. Each of the links was in a constant state of flux, so there could be no permanent Self or soul that survived death. Nevertheless, as a cycle powered by kamma, dependent origination could keep on turning and somehow span

different realms of existence ad infinitum. This was pure speculation, however, because while one could test the operation of the
cycle against one's own experience of suffering in this life, what
happened to "you" after death was anyone's guess.

To his credit, Siddhārtha actively discouraged his followers
from indulging in too much speculation about imponderables such
as the origin of the cosmos, what happens to a fully enlightened
being when he or she dies, and the precise operation of kamma.[9]
He insisted that the objective of his teaching was to prevent suffering in *this* life. Everything else was a sideshow. He compared the
kind of person who raises endless metaphysical questions to a man
struck by an arrow "thickly smeared with poison" who refuses to
have it extracted until he knows whether the person who fired it
was a warrior, a brahmin, a merchant, or a worker . . . his name,
how tall he was, his complexion and his hometown . . . the materials used to make the bow and arrow . . . "The man would die," the
Buddha concluded, "and those things would still remain unknown
to him."[10] The poisoned arrow represented human suffering and
the sole purpose of the Buddha's teaching was to extract it. Rather
than allowing themselves to get caught up in metaphysical specu-
lations, he advised his followers to reflect continually:[11]

> "I am subject to aging, have not gone beyond aging . . .
> "I am subject to illness, have not gone beyond
> illness . . .
> "I am subject to death, have not gone beyond
> death . . .
> "I will grow different, separate from all that is dear
> and appealing to me . . .

"I am the owner of my kamma, heir to my kamma,
born of my kamma, related through my kamma, and have
my kamma as my arbitrator. Whatever I do, for good or
for evil, to that will I fall heir."

None of these reflections should come as any surprise to modern people—we are perfectly aware on one level that nothing lasts forever, that our bodies will wear out and we will die, and that we must face up to the consequences of our actions—and yet there is a tenacious part of our brain that denies them. The neuroscientific and clinical evidence points the finger of blame for this mental blind spot at the default mode network and its simulation of our overriding sense of unchanging selfhood: the avatar that flits backward and forward through time labeling transient stuff such as our bodies and every fleeting thought, feeling, and emotion as "I, me, and mine." This Self makes us feel worthy of immortality, so even though we came into the world with nothing, we feel cruelly cheated by death. A passage in the Dhammapada sums up our distress at the prospect of being deprived of the things that were never really ours in the first place: [12]

This body is a painted image, subject to disease, decay
and death, held together by thoughts that come and go.
What joy can there be for those who see their white bones
will be cast away like gourds in the autumn?

But a few verses later come the defiant words quoted at the start of this chapter, which Siddhārtha is said to have uttered after attaining enlightenment: [13]

I have gone through many rounds of birth and death,
looking in vain for the builder of this body. Heavy indeed
is birth and death again and again! But now I have seen
you, housebuilder; you shall not build this house again.
Its beams are broken; its dome is shattered: self-will is ex-
tinguished; nirvana is attained.

After six years of practicing meditation and self-restraint, he finally saw through the artifices of his brain and awakened to reality.
Nothing was permanent: whatever arose in his mind would eventually cease. Ignorance dispelled, free of all craving and aversion, he
had broken the chain of dependent origination and brought the cycle
of birth, death, and suffering to an end. For a practicing Buddhist, to
realize fully the impermanent, nonpersonal nature of all mental and
physical phenomena is the ultimate fulfillment of the possibilities
of mindfulness, the final destination on their spiritual journey—the
Deathless Realm from which there is no return.[14] Simply paying attention, dispassionately, to what is happening in the present moment
is said to break the chain. By dispelling once and for all any remaining delusions about the true nature of existence, mindfulness will
eventually end the cycle of birth, death, and rebirth that perpetuates
suffering. As the Buddha said, according to the Dhammapada:[15]

Mindfulness is the path to the deathless
Heedlessness is the path to death.
The mindful do not die;
But the heedless are as if dead already.

To make this journey toward the final enlightenment isn't a
trivial undertaking. It is said to take years of dedicated practice

and most of the time the contemplative must walk the path alone: even the greatest guru can't take you to the Deathless Realm. And because there are no words to explain adequately the associated state of mind, for the vast majority of us who have not experienced it directly, imagination fails us. Buddhist texts describe the most refined levels of meditation preceding perfect enlightenment— equivalent to the four loftiest "formless realms" of existence—as the progressive realization of infinite space, then infinite consciousness, then nothingness, and finally "neither perception nor non-perception."[16] When the words describing these sublime states of being are so slippery, it is easy to become skeptical about whether they exist. Having said that, even people who have never practiced meditation can get an inkling of its long-term possibilities when they first learn to focus on their breath and discover that their mind becomes brighter, calmer, and less self-absorbed as its default mode of operation falls away. Who knows what might be possible after decades of dedicated, intensive practice?

Preliminary evidence from electroencephalography (EEG) studies suggests that something extraordinary really is happening in the brains of contemplatives who report experiencing the most rarefied states of consciousness. Richard Davidson, Antoine Lutz, and their colleagues at the University of Wisconsin applied electrodes to the scalps of eight Tibetan monks who had each accumulated between ten thousand and fifty thousand hours of meditation over periods ranging from fifteen to forty years.[17] The scientists then recorded the electrical activity in their brains as they practiced an "objectless" form of meditation described as a state of "unrestricted readiness and availability to help all living beings." Rather than concentrating on a memory, an image, or a sensation such as the breath, the meditator fills his or her mind with pure or nonreferential compassion. The researchers found that this state was associated with intense "gamma rhythms" in their brains. We know that when vast assemblies of neurons fire in unison, this creates rhythmic electrical oscillations that can spread across the brain, allowing remote regions to synchronize their activity. More ordinary meditative states created by focusing on the breath, for example, are associated with synchronous, slow electrical rhythms known as alpha and theta rhythms, with frequencies less than 15Hz, but when the contemplatives in Lutz and Davidson's study practiced objectless meditation they generated gamma rhythms—strong, fast oscillations with a frequency greater than 25Hz. The researchers report that the strength or "amplitude" of these rhythms was greater than any previously seen in healthy individuals. Compared with ten control subjects, who had been practicing meditation for only a week, the monks also had stronger gamma rhythms during a resting state when they were not meditating, which suggests that their years of practice had created lasting changes in the way their brains worked. The more hours of meditation they had notched up, the greater the intensity of gamma activity recorded.

The exact function of gamma oscillations in the brain is unknown, but they appear to be involved in information processing of some sort. Less intense gamma oscillations are associated with waking consciousness, hallucinations, and rapid eye movement (REM) sleep, when vivid dreaming occurs. Recent research also links them, albeit indirectly, to the phenomenon of "near death experiences." People who have been brought back from the brink of death by doctors sometimes describe vivid hallucinations, a bright light, the sensation of levitation or looking down on their own body, and intensely pleasant feelings such as serenity and security, unconditional love and acceptance. In one study, around

20 percent of people who were successfully revived after cardiac arrest reported having such an experience.[18] The brain activity of humans shortly after cardiac arrest has never been recorded using EEG, but in 2013, scientists at the University of Michigan detected a strong burst of gamma synchrony in the brains of rats within thirty seconds of an induced heart attack.[19] The intensity of the gamma rhythms was greater even than that seen during the animals' normal waking state. It is impossible to know whether the animals were having the rodent equivalent of a near death experience while this was happening, but the discovery that activity in a mammal's brain can briefly *increase* after blood flow has ceased suggests that in humans these experiences are physiological rather than supernatural in origin. The research also raises the intriguing possibility that when highly accomplished contemplatives induce feelings such as boundless, unconditional love or compassion during objectless meditation, their state of mind is similar to that of someone having a near death experience. By the same token, intense gamma rhythms in the brain may also be responsible for other peculiar things that religious people down the ages have experienced during profound contemplative states, such as vivid hallucinations, a blinding light, and the sensation of levitating.

There is another sound reason to give credence to Buddhist contemplatives' reports of enlightenment experiences. The total integration of mindfulness into the flow of everyday life is reputed to liberate one from fear of death (though it is important to note that the ultimate objective of mindfulness as conceived in Buddhism is not to reject life but rather to relinquish all attachments to what is impermanent). Something very similar appears to happen quite naturally for some people diagnosed with a terminal illness. By coming to terms with their own imminent death, they let go of the things they formerly clung to in life. Paradoxically, every experience then becomes more vivid, and they report that they never felt so alive.

His fists permanently bunched as a result of a painful condition called psoriatic arthropathy that had afflicted him for decades, British television dramatist Dennis Potter somehow managed to grip a lit cigarette in one fist and take occasional sips from a glass of white wine clenched by its stem in the other. He was being interviewed by broadcaster Melvyn Bragg in March 1994 for Channel 4 in the UK.[20] A month earlier he had been diagnosed with pancreatic cancer, which had spread to his liver and was inoperable. He told Bragg that his one remaining objective was to finish the play he was working on, so he had asked his doctor to limit his pain relief medication to allow him to remain in full control of his faculties in these final months of his life. Just once during the interview Potter took a swig from a hip flask of morphine to control a painful spasm. He remained perfectly lucid throughout. He was concerned about the effect of his imminent death on friends and family, he said, but he hadn't experienced a single moment of terror on his own account. He explained to Bragg that his diagnosis had led to an important revelation. "We're the one animal that knows we're going to die and yet we carry on paying our mortgages, doing our jobs, moving about, behaving as though there's eternity, and we forget that life can only be defined in the present tense. It is 'is' and it is now only," he said. "That 'nowness' has becomes so vivid to me I'm almost serene. I can celebrate life." As an example he described a plum tree in full bloom beneath the window of the room where he sometimes wrote. "It is the whitest, frothiest, *blossomest* blossom that there could ever be, and I can see it. And things are both more trivial than they ever were and more important than they ever were, and the difference between the trivial and the important doesn't seem to matter. But the nowness of everything is absolutely wonderful and if people could *see* that . . . there's no way of telling you, you have to experience it. But the glory of it, the comfort of it, the reassurance of it—not that I'm interested in reassuring people, bugger that!—the fact is that if you see the present tense, boy do you see it. And boy can you celebrate it."

In some people the imminence of death is strangely liberating. Potter met his final deadline, finishing his play *Cold Lazarus* before he died on June 7, 1994. Regardless of whether he even knew the word, mindfulness seems to have helped him discover a sense of joy and serenity in his final months. Many others in the same position have offered similar reports: a feeling of peace and heightened intensity of experience in the present moment that seems to arise from total acceptance of the transience of life. There is an obvious lesson here. Even the ultrarich, with their impregnable panic rooms and platinum health insurance plans, can't escape sickness, old age, and death, but the more we *have*, the more tenaciously we cling to our transient youth and health, our possessions, and to life itself. This clinging—or "attachment" as a Buddhist would call it—may actually offset the potential of wealth and security to make us happier, more fulfilled people. If so, that would go a long way toward explaining the "happiness paradox"—the finding that self-reported happiness ratings peaked in the 1950s in developed countries such as the UK and US and have changed little since, despite steep rises in GDP in the intervening years.[21, 22] In material terms, the benefits of prosperity and technology are obvious. To a man or woman living in the fifth century BCE, sentient beings who knew they would always have enough to eat, who could fly, set foot on the moon, send probes to explore the solar system, and communicate effortlessly with others on the far side of the Earth would have seemed like gods. But despite ascending to these heavenly realms, the flawed human mind hasn't changed since the Iron Age. It clings to the delusions of permanence, against all the evidence, and this still causes us enormous suffering. According to the legend, before Siddhārtha's enlightenment, even the gods, angels, and devas from the higher realms of existence were looking to him for release from the endless cycle of death, rebirth, and death. Like spectators flocking to a prize fight, they came to watch his bid to win enlightenment that evening beneath the bodhi tree, though they fled in terror when his opponent Mara—"Death"—and his army arrived at the venue. Siddhārtha defeated Death, though not in any literal sense. You could call it a moral victory.

In the ancient Indian language Pāli, *Amaravati* means the Deathless Realm. It is not only the name of a very modern monastery near Hemel Hempstead in the United Kingdom where Ajahn Amaro is the abbot, but also a village in Andhra Pradesh, India, site of the remains of a magnificent stupa dating from the third century BCE that was once richly decorated with beautiful sculptures. One of these, now on display at the Guimet Museum of Asian Arts in Paris, depicts Mara and his fearsome army menacing the Buddha beneath the bodhi tree. Several other sculptural scenes recovered from the site incorporate the cartwheel-like Dharmachakra, the Buddhist wheel of moral law. One particularly fine example of Amaravati sculpture on display at the British Museum in London depicts Emperor Ashoka with his hands pressed reverently together, flanked by attendants and his queen. We would not know who this man was or his devotion to the Dhamma were it not for the edicts that he ordered to be chiseled into rocks and pillars across his vast empire. In the nineteenth century, archaeologists identified a total of twenty-nine rock edicts and nine pillar edicts, one of which was found in fragments among the ruins of the stupa at Amaravati. Many of the emperor's lengthy, often rambling directives have been translated. In the words of the historian Charles Allen, they reveal a man "deeply—even obsessively—spiritual, passionate in his belief in a higher morality, in showing kindness and helping the poor, in moderation and self-control, in tolerance for all religions, in the sanctity of life, in the virtues of self-examination, truthfulness, purity of heart and, above all, in his love of the Dharma."[23] Upturning every convention of how an all-powerful monarch should behave, Ashoka ordered it to be set in stone that he no longer believed in conquest by military force. If there was to be conquest, it would be by Dhamma alone.

Ashoka's example points to a modern interpretation of the operation of kamma that spans generations, but without any need to invoke the concept of rebirth. The idea that the good actions and intentions of individuals can affect not only their own future well-being but also that of those around them and future generations is just as meaningful regardless of whether one believes in the continuity of life after death. In a very real sense, the accumulated, *collective kamma* of our ancestors—including the Emperor Ashoka, who in all probability saved Buddhism from obscurity—has carried us to "the realms of the gods." Over the course of thousands of years and hundreds of generations, human society has advanced technologically, culturally, morally. On the whole, suffering has been reduced and well-being improved. Among many other things, we have our ancestors to thank for agriculture, writing, arithmetic, medicine, education, democracy, the rule of law, and the concept of universal human rights. If research collated by the psychologist Steven Pinker in his book *The Better Angels of Our Nature* is to be accepted, overall levels of violence and conflict have declined markedly over the past two millennia.[24] These cumulative advances have been won not as a result of supernatural forces, let alone through the operation of "selfish genes," but by the selfless actions of individuals, both the celebrated and the anonymous, the great and the small. The Buddha and the emperor played their part, but so have billions of others, the plumbers and politicians, artists and scientists, mothers and fathers, brothers and sisters, friends and caregivers. This is kamma on a grand scale, a steady accretion of collective good intentions and deeds that has helped humanity transcend its savage origins. Buddhism teaches that the distinction between ourselves and others is an artificial one born from the delusion of selfhood. If we can only learn to let go of this Self, perhaps the distinction between humans still living, those who have died, and those yet unborn will no longer seem so important.

Of course, not all is well with us. After all, our basic nature is still that of any other top predator. Our unchecked cravings for pleasure and status are depleting the world's natural resources, aversion drives conflict, delusion blinds us. Focusing on the example most relevant to this book, persistently high rates of mental illness worldwide—even in the richest nations—suggests that much work remains to be done to optimize our collective kamma. I have argued that things started to go awry for the human brain on the African savannah after our evolutionary line split from that of chimps some six million years ago. According to the social brain hypothesis developed by psychologist Robin Dunbar, in order to survive in this dangerous environment, our ancestors began to band together in ever larger groups. Their brains evolved a sophisticated theory of mind that allowed them to navigate their increasingly complex social world by intuiting the beliefs and intentions of numerous others. Out of this ability to model the minds of others arose the sense of Self as the central player in this social drama, and to keep track of its trajectory in relation to all the other players, our minds ranged deeper into the past and the future, reconstructing past scenarios and imagining future ones, musing on what had happened and planning what to do next, going far beyond the mental-time-traveling and planning abilities of any other primate. In step with these changes, the development of grammatical spoken language from simple manual gestures over the course of millions of years allowed us to share with others not only our private mental time travels but also our experiences in the real world. We became very good at learning from one another, accumulating collective knowledge, skills, and wisdom that could be passed between tribes and down the generations— collective good kamma, if you will.

These novel mental abilities had three significant costs. First, the mind's newfound ability to simulate the future and the past meant it could now wander further and further from the realities of the present moment and the plain evidence of our senses. Secondly, as a result of our extraordinary faculty for modeling the minds of others in our own, we became masters of deception, and more suspicious of the intentions of others. Finally, the upstart Self now asserted ownership rights over other people, physical objects, and experiences, plastering them with sticky "I, me, mine" labels that would prove painful to remove when the time came, which it invariably did.

Neuroscientists have discovered that the brain's default mode network plays a central role in all our newly evolved capabilities and their associated downsides. It not only facilitates mental time travel, theory of mind, and the sense of Self, but its misfirings have been implicated in mental illnesses such as depression and anxiety. The scientific research reviewed in this book provides evidence that meditation can address some of the weaknesses of our time-traveling, meandering, egotistical social brains. Practicing mindfulness can quieten the default mode network, helping us to focus on the present; it sharpens our powers of attention and metacognition; and by weakening the grip of the Self makes it easier to "let go." Mindfulness is not a panacea, but for those who are willing to put in the time and effort, evidence suggests that it can achieve remarkable things. In the clinical sphere, it has been shown to prevent depression, reduce anxiety, tackle substance-abuse cravings, and ease pain. And as psychiatrists come to the realization that the sharp distinction they once drew between the mentally well and ill is an artificial one, you could call the universal benefits of mindfulness "optimal mental well-being." Or you could call them "enlightenment," it doesn't really matter.

To help his followers along this path, the Buddha taught Seven Factors of Enlightenment,[25] which also serve as a short summary of the topics addressed within the pages of this book. The more closely I look at them, the more it becomes obvious that they are equally applicable to those who seek enlightenment through scientific research, or indeed any other field of endeavor:

Calm. Prepare for the task ahead by relaxing body and mind. This will clear the fog of old emotional habits and preconceptions so you can see the task ahead more clearly.

Concentration. Cultivate the ability to apply sustained, focused attention. This will help you avoid all kinds of carelessness and error.

Mindfulness. Observe all mental and physical phenomena with objectivity from moment to moment. Put your ego to one side.

Investigation. Experiment. When such and such happens, what are the consequences?

Energy. Single-minded determination will be needed to see your project through.

Equanimity. Treat pleasure and pain, praise and blame, triumph and disaster just the same. Try not to take everything so personally.

Happiness or rapture. A happy, contented mind is an effective mind, helping you along your path. There may even be eureka moments (but see *equanimity*, above).

Practiced with compassion and allied to the highest ethical standards, Buddhists believe that cultivating these qualities brings one closer to enlightenment. Of course, other world religions have reached similar conclusions. Uniquely, however, Buddhism has never imposed a creed or demanded that its adherents believe in any god. What's more, one needn't be a Buddhist to benefit from the insights into the human mind of its founder Siddhārtha Gautama. One need only start practicing.

Even perfectly enlightened beings are mortal. At the age of eighty, the Buddha fell ill with food poisoning while traveling with his faithful attendant Ananda and a large company of monks through a remote region in northeast India, far from any city. "I am now grown old and full of years," he said. "My journey is drawing to its close, I have reached the sum of my days." Ananda was distraught that his master was dying. "Enough, Ananda! Do not grieve, do not lament!" Hadn't he taught him that nothing was permanent, that the wise do not grieve, having realized the nature of existence? In a grove near the town of Kusinara, Ananda made a bed for him between two sala trees.[26] Almost to the end, the Buddha continued to receive visitors and teach. "Be lamps unto yourselves," he exhorted them. "Rely on yourselves, do not rely on external help. Hold fast to the truth as a lamp. Seek salvation alone in the truth."

When the time came, he settled into his final meditation. Forty-five years earlier he had relinquished the man who was Prince Siddhārtha Gautama, son of King Suddhodana of Shakya. Now he began to let go of everything else: his body, thoughts, feelings, senses. Rising through increasingly diffuse realms of consciousness into infinite space and nothingness, his brain expired in an ecstatic burst of high-frequency electrical oscillations.

ACKNOWLEDGMENTS

This book could not have been written without the generosity of the neuroscientists, psychologists, and psychiatrists who cleared spaces in their busy schedules for interviews to describe for me their adventures in the uncharted territory where the brain, the mind, and well-being come together. I'd like to extend my heartfelt thanks to Jon Kabat-Zinn and Judson Brewer at the University of Massachusetts Medical School; Richard Davidson at the University of Wisconsin–Madison; Herbert Benson at Massachusetts General Hospital; Zindel Segal at the University of Toronto; Willem Kuyken and Daniel Freeman at the University of Oxford; Britta Hölzel at the Technische Universität München; Simon Wessely at the Royal College of Psychiatrists; David Haslam at NICE; Sarah Bowen at the University of Washington; and Judith Soulsby at the University of Bangor.

I also owe a huge debt of gratitude to the monks, nuns, and lay staff of Amaravati Buddhist Monastery for the warmth of their hospitality, and in particular to the abbot Ajahn Amaro, whose wise words remain an inspiration. Amaravati embodies an ancient Eastern philosophy that runs counter to almost everything Westerners are brought up to believe about the pursuit of happiness. That this place and its philosophy are thriving in the twenty-first century was a source of wonder to me.

Siddhārtha's Brain is a product of that wonder. I'd like to thank my agent, Peter Tallack, for believing in this strange hybrid of spirituality, neuroscience, and psychiatry when it was no more than a sketchy idea, and for his and Tisse Takagi's sterling work transforming it into a viable proposal. It caught the eye of the editor Peter Hubbard at William Morrow/HarperCollins, who was brave or crazy enough to take a gamble on this first-time author. I have him and his excellent team of copy editors and designers to thank for the book you have in your hands or on your screen. I am grateful also to Louisa Pritchard for her tireless work selling the rights to translations around the world.

Finally, I have my lovely family and friends to thank for their unfailing support, especially in the early stages of the project when they probably assumed some kind of midlife crisis was under way. In particular I would like to thank Charlotte for rescuing me when, for a few days, it really did look as though things were starting to unravel. And I will be forever indebted to Art, who did more than anyone else to set my feet on this path.

NOTES

All excerpts of scriptural translations, unless otherwise stated, are from www.accesstoinsight.org.

INTRODUCTION

1. Harrington, A. and Zajonc, A. (eds.), *The Dalai Lama at MIT* (Harvard University Press, 2008), p. 63.
2. Yee, C. M., Javitt, D. C. and Miller, G. A. (2015), "Replacing *DSM* Categorical Analyses with Dimensional Analyses in Psychiatry Research: The Research Domain Criteria Initiative," *JAMA Psychiatry* 72(12): 1159–60.
3. Adam, D. (2013), "Mental Health: On the Spectrum," *Nature* 496: 416–18.
4. Ronald, A. et al. (2013), "Characterization of Psychotic Experiences in Adolescence Using the Specific Psychotic Experiences Questionnaire: Findings from a Study of 5,000 16-Year-Old Twins," *Schizophrenia Bulletin* (doi: 10.1093/schbul/sbt106).
5. Freeman, D. et al. (2011), "Concomitants of Paranoia in the General Population," *Psychological Medicine* 41(5): 923–36.
6. Freeman, D. et al. (2008), "Virtual Reality Study of Paranoid Thinking in the General Population," *British Journal of Psychiatry* 192: 258–63.
7. Ohayon, M. M. (2000), "Prevalence of Hallucinations and Their Pathological Associations in the General Population," *Psychiatry Research* 97(2–3): 153–64.
8. Jones, S. et al., *Understanding Bipolar Disorder* (The British Psychological Society, 2010).
9. Helliwell, J., Layard, R. and Sachs, J., *World Happiness Report 2013*.
10. Pedersen, Carsten Bøcker, et al. (2014), "A Comprehensive Nationwide Study of the Incidence Rate and Lifetime Risk for Treated Mental Disorders," *JAMA Psychiatry* 71(5): 573–81.

11. *Global Burden of Diseases, Injuries and Risk Factors Study 2010* (Institute for Health, Metrics and Evaluation, 2010).
12. Layard, R. et al. (2014), "What Predicts a Successful Life? A Life-Course Model of Well-Being," *The Economic Journal* 124 (F720–38).
13. Election broadcast by the Natural Law Party, https://www.youtube.com/watch?v=438UKM1Av1g.
14. Fox, K. C. R. et al. (2014), "Is Meditation Associated with Altered Brain Structure?: A Systematic Review and Meta-Analysis of Morphometric Neuroimaging in Meditation Practitioners," *Neuroscience and Behavioral Reviews* 43: 48–73.
15. Davidson, R. J. and Schuyler, B. S., "Neuroscience of Happiness," in: *World Happiness Report 2015*.
16. Kingsland, J., "The Rise and Fall of the Wonder-Drugs," *New Scientist*, July 3, 2004.
17. Gibbons, R. D. et al. (2012), "Benefits from Antidepressants: Synthesis of 6-Week Patient-Level Outcomes from Double-Blind Placebo-Controlled Randomized Trials of Fluoxetine and Venlaflaxine," *Archives of General Psychiatry* 69(6): 572–79.
18. Kirsch, I. et al. (2008), "Initial Severity and Antidepressant Benefits: A Meta-Analysis of Data Submitted to the Food and Drug Administration," *PLOS Medicine* 5(2): 260–68.
19. Fournier, J. C. et al. (2009), "Antidepressant Drug Effects and Depression Severity: a Patient-Level Meta-Analysis," *JAMA* 303(1): 47–53.

CHAPTER 1: A FOOL'S PARADISE

1. Majjhima Nikāya, 12, 36.
2. Anguttara Nikāya, i 145.
3. Nidana Katha (*The Story of the Lineage*), translated by Rhys Davids, T. W. (Routledge, 1925).
4. Majjhima Nikāya, 36.
5. Jackson, T., *Prosperity Without Growth* (Earthscan, 2011), p. 40.
6. Helliwell, J., Layard, R. and Sachs, J., *World Happiness Report 2013*, chap. 3.
7. *Investing in Mental Health* (World Health Organization, 2003).
8. *Adult Psychiatric Morbidity in England—2007* (Health & Social Care Information Centre, 2007).
9. World Health Organization Secretariat (2011), "Global Burden of Mental Disorders and the Need for a Comprehensive, Coordinated Response from Health and Social Sectors at the Country Level."
10. Tsai, A. C. and Tomlinson, M. (2015), "Inequitable and Ineffective: Exclusion of Mental Health from the Post-2015 Development Agenda," *PLOS Medicine* 12(6): e1001846.
11. Ferreira, A. et al. (2011), "Sickle Hemoglobin Confers Tolerance to *Plasmodium* Infection," *Cell* 145(3): 398–409.

12. Asimov, I., *Asimov's New Guide to Science* (Penguin Books, 1987).
13. Power, R. A. et al. (2013), "Fecundity of Patients with Schizophrenia, Autism, Bipolar Disorder, Depression, Anorexia Nervosa, or Substance Abuse vs Their Unaffected Siblings," *JAMA Psychiatry* 70(1): 22–30.
14. Kivimäki, M. et al. (2012), "Job Strain As a Risk Factor for Coronary Heart Disease," *The Lancet* 380(9852): 1491–97.
15. Bonde, J. P. E. (2008), "Psychosocial Factors at Work and Risk of Depression: A Systematic Review of the Epidemiological Evidence," *Occupational & Environmental Medicine* 65: 438–45.
16. Sprong, M., Scothorst, P. and Vos, E. (2007), "Theory of Mind in Schizophrenia: Meta-Analysis," *British Journal of Psychiatry* 191: 5–13.
17. Killingsworth, M. A. and Gilbert, D. T. (2010), "A Wandering Mind Is an Unhappy Mind," *Science* 330: 932.
18. Lennon, J., "Beautiful Boy (Darling Boy)," *Double Fantasy* (1980).
19. The Dhammapada, translated by Easwaran, E. (Nilgiri, 2007), verses 1–2.
20. Anguttara Nikāya, 3.65.
21. The Dhammapada, translated by Byrom, T., (Shambhala, 1993), verse 80.
22. Goyal, M. et al. (2014), "Meditation Programs for Psychological Stress and Well-Being: A Systematic Review and Meta-Analysis," *JAMA Internal Medicine* 174(3): 357–68.
23. Piet, J. and Hougaard, E. (2011), "The Effect of Mindfulness-Based Cognitive Therapy for Prevention of Relapse in Recurrent Major Depressive Disorder: A Systematic Review and Meta-Analysis," *Clinical Psychology Review* 31(6): 1032–40.
24. Kuyken, W. et al. (2015), "Effectiveness and Cost-Effectiveness of Mindfulness-Based Cognitive Therapy Compared with Maintenance Anti-Depressant Treatment in the Prevention of Depressive Relapse/Recurrence: Results of the PREVENT Randomised Controlled Trial," *The Lancet* 386: 63–73.
25. Goyal, M. et al. (2014), "Meditation Programs for Psychological Stress and Well-Being: A Systematic Review and Meta-Analysis," *JAMA Internal Medicine* 174(3): 357–68.
26. Armstrong, Karen, *Buddha* (Phoenix, 2000).
27. Majjhima Nikāya, 12.

CHAPTER 2: CHILD'S PLAY

1. Majjhima Nikāya, 36.
2. Benson, H. et al. (1969), "Behavioral Induction of Arterial Hypertension and Its Reversal," *American Journal of Physiology* 217: 30–34.
3. Benson, H. and Klipper, M. Z., *The Relaxation Response* (HarperCollins, 1975).
4. Cannon, W. B., *Bodily Changes in Pain, Hunger, Fear and Rage: An Ac-*

count of Recent Researches into the Function of Emotional Excitement (Appleton, 1915).

5. Kozinn, A., "Meditation on the Man Who Saved the Beatles," *New York Times,* February 7, 2008.

6. Miles, B. *Paul McCartney: Many Years From Now* (Henry Holt and Company, 1997).

7. Nidich, S. I., et al. (2009), "A Randomized Controlled Trial on Effects of the Transcendental Meditation Program on Blood Pressure, Psychological Distress, and Coping in Young Adults," *American Journal of Hypertension* 22(12): 1326–31.

8. AHA Scientific Statement (2013), "Beyond Medications and Diet: Alternative Approaches to Lowering Blood Pressure," *Hypertension* 61: 1360–83.

9. Dusek, J. et al. (2008), "Genomic Counter-Stress Changes Induced by the Relaxation Response," *PLOS ONE* 3: e2576.

10. Bhasin, M. K. et al. (2013), "Relaxation Response Induces Temporal Transcriptome Changes in Energy Metabolism, Insulin Secretion and Inflammatory Pathways," *PLOS ONE* 8: e62817.

11. Cawthon, R. et al. (2003), "Association Between Telomere Length in Blood and Mortality in People Aged 60 Years or Older," *The Lancet* 361: 393–95.

12. Epel, E. S. et al. (2004), Accelerated Telomere Shortening in Response to Life Stress," *Proceedings of the National Academy of Sciences* 101 (49): 17312–15.

13. Epel, E. S. et al. (2006), "Cell Aging in Relation to Stress Arousal and Cardiovascular Disease Risk Factors," *Psychoneuroendocrinology* 31: 277–87.

14. Jacobs, T. L. et al. (2011), "Intensive Meditation Training, Immune Cell Telomerase Activity, and Psychological Mediators," *Psychoneuroendocrinology* 36: 664–81.

15. Majjhima Nikāya, 36.

16. Majjhima Nikāya, 46.

17. *Buddhist Birth-Stories (Jataka Tales), Nidāna-Kathā: The Story of the Lineage,* adapted from the translation by Rhys Davids, T. W. (Routledge, 1925).

CHAPTER 3: THE CLOUD OF UNKNOWING

1. Narada Maha Thera, *The Buddha and His Teachings* (Buddhist Publication Society, 1988).

2. Possehi, G. L., *The Indus Civilization: A Contemporary Perspective* (AltaMira Press, 2002).

3. Taimni, I. K., *The Science of Yoga: The Yoga Sutras of Patanjali* (Quest Books, 1999).

4. The Bhagavad Gita, translated by Easwaran, E. (Nilgiri Press, 2007).

5. Taimni, I. K., *The Science of Yoga: The Yoga Sutras of Patanjali* (Quest Books, 1999).

6. Dhamma talk by Ajahn Chah delivered to newly ordained monks at Wat Nong Pah Pong, Thailand, in July 1978. www.ajahnchah.org.

7. Anonymous, *The Cloud of Unknowing*, translated by Spearing, A. C. (Penguin Classics, 2001).

8. de Osuna, F., *The Third Spiritual Alphabet* (Benziger Brothers, 1931).

9. Scholem, G., *Major Trends in Jewish Mysticism* (Schocken Books, 1996).

10. Kelly, E. M., *The Rosary: A Path into Prayer* (Loyola University Press, 2004).

11. Reynold Nicholson, A., *The Mystics of Islam* (World Wisdom, 2003).

12. "The Sufis and St Francis of Assisi," in: Shah, I., *The Sufis* (Doubleday, 1964).

13. Gethin, R., *The Foundations of Buddhism* (Oxford University Press, 1998).

14. "Development of Mahamudra," in: *Teachings of the Buddha*, edited by Jack Kornfield (Shambhala Publications, 1996).

CHAPTER 4: THE SECOND DART

1. Vinaya Mahāvagga, 1:6.

2. Saṃyutta Nikāya, 56:11.

3. Majjhima Nikāya, 26.

4. "The Four Noble Truths," in: Ajahn Sumedho, *Peace Is a Simple Step* (Amaravati Publications, 2014).

5. Saṃyutta Nikāya, translated by Nyanaponika Thera, 36:6.

6. The Health and Social Care Information Centre, *Health Survey for England—2011*, chap. 9, "Chronic Pain."

7. Institute of Medicine of the National Academies, *Relieving Pain in America* (The National Academies Press, 2011).

8. Kabat-Zinn, J. (1982), "An Outpatient Program in Behavioral Medicine for Chronic Pain Patients Based on the Practice of Mindfulness Meditation: Theoretical Considerations and Preliminary Results," *General Hospital Psychiatry* 4: 33–47.

9. Goyal, M. et al. (2014), "Meditation Programs for Psychological Stress and Well-Being: A Systematic Review and Meta-Analysis," *JAMA Internal Medicine* 174(3): 357–68.

10. Chen, K. W. et al. (2012), "Meditative Therapies for Reducing Anxiety: A Systematic Review and Meta-Analysis of Randomized Controlled Trials," *Depression and Anxiety* 29: 545–62 (doi: 10.1002/da.21964 PMID: 22700446).

11. Kabat-Zinn, J., Lipworth, L. and Burney, R. (1985), "The Clinical Use

of Mindfulness Meditation for the Self-Regulation of Chronic Pain," *Journal of Behavioral Medicine* 8(2): 163–90.

12. Morone, N. et al. (2008), "Mindfulness Meditation for the Treatment of Chronic Low Back Pain in Older Adults: A Randomized Controlled Pilot Study," *Pain* 134(3): 310–319.

13. Rosenzweig, S. et al. (2010), "Mindfulness-Based Stress Reduction for Chronic Pain Conditions: Variation in Treatment Outcomes and Role of Home Meditation Practice," *Journal of Psychosomatic Research* 68: 29–36.

14. Gaylord, S. A. et al. (2011), "Mindfulness Training Reduces the Severity of Irritable Bowel Syndrome in Women: Results of a Randomized Controlled Trial," *The American Journal of Gastroenterology* 106: 1678–88.

15. Grossman, P. et al. (2007), "Mindfulness Training as an Intervention for Fibromyalgia: Evidence of Postintervention and 3-Year Follow-Up Benefits in Well-Being," *Psychotherapy and Psychosomatics* 76: 226–33.

16. Zeidan, F. et al. (2011), "Brain Mechanisms Supporting the Modulation of Pain by Mindfulness Meditation," *The Journal of Neuroscience* 31(14): 5540–48.

17. Craig, A. D. (2009), "How Do You Feel—Now? The Anterior Insula and Human Awareness," *Nature Reviews: Neuroscience* 10: 59–70.

18. Grant, J. A. et al. (2011), "A Non-Elaborative Mental Stance and Decoupling of Executive and Pain-Related Cortices Predicts Low Pain Sensitivity in Zen Meditators," *Pain* 152(1): 150–56.

19. Grant, J. A. et al. (2010), "Cortical Thickness and Pain Sensitivity in Zen Meditators," *Emotion* 10(1): 43–53.

20. Gard, T. et al. (2012), "Pain Attenuation through Mindfulness Is Associated with Decreased Cognitive Control and Increased Sensory Processing in the Brain," *Cerebral Cortex* 22: 2692–2702.

21. Wiech, K. et al. (2008), "Neurocognitive Aspects of Pain Perception," *Trends in Cognitive Science* 12: 306–13.

22. Zeidan, F. et al. (2010), "The Effects of Brief Mindfulness Meditation Training on Experimentally Induced Pain," *The Journal of Pain* 11(3): 199–209.

23. Wager, T. D. et al. (2004), "Placebo-Induced Changes in fMRI in the Anticipation and Experience of Pain," *Science* 303: 1162–67.

24. Zeidan, F. et al. (2015), "Mindfulness-Based Pain Relief Employs Different Neural Mechanisms than Placebo and Sham Mindfulness Meditation-Induced Analgesia," *The Journal of Neuroscience* 35(46): 15307–325.

25. The Bhagavad Gita, translated by Easwaran, E. (Nilgiri Press, 2007), 2: 14–15.

26. Ibid., 2: 62–63.

27. Bronkhorst, J., *The Two Traditions of Meditation in Ancient India* (Motilal Banarsidass, 1993), p. 70.

28. Saṃyutta Nikāya, 56.11.

CHAPTER 5: THE MAN WHO DISAPPEARED

1. Saṃyutta Nikāya, 22:59.
2. Vinaya Mahāvagga, 1:6.
3. Rig Veda, X.97.11.
4. Brihadaranyaka Upanishad, translated by Swāmi Madhavananda (Swami Yogeshwarananda, 1950), 3.7.23.
5. *Pausanias: Description of Greece, Volume IV,* translated by Jones, W. H. S. (Harvard University Press, 1935), book 10, chap. 24.
6. Exodus 3:14 (AV).
7. Pagels, E., *The Gnostic Gospels* (Vintage, 1989), p. xix.
8. Frager, R., *Heart, Self, and Soul: A Sufi Approach to Growth, Balance, and Harmony* (Quest Books, 1999).
9. The Bhagavad Gita, translated by Easwaran, E. (Nilgiri Press, 2007), 6: 19–20, 26–27.
10. Anonymous, *The Cloud of Unknowing,* translated by Spearing, A. C. (Penguin Classics, 2001).
11. The Dhammapada, translated by Easwaran, E. (The Blue Mountain Center of Meditation, 2007), pp. 82–83.
12. Hume, D., *A Treatise of Human Nature* (Oxford University Press, 2011).
13. Vinaya Mahāvagga, 1:6.
14. Monod, J., *Chance and Necessity,* translated by Wainhouse, A. (Fontana, 1974).
15. Parfit, D., *Reasons and Persons* (Oxford Paperbacks, 1986).
16. Bogen, J. E., "The Callosal Syndrome," in: Heilman, K. M. and Valenstein, E. V. (eds.), *Clinical Neuropsychology*: 295–338 (Oxford University Press, 1979).
17. Gazzaniga, M. S. (1967), "The Split Brain in Man," *Scientific American* 217: 24–29.
18. Eagleman, D., *Incognito: The Secret Lives of the Brain* (Canongate Books, 2011).
19. Kelley, W. M. et al. (2002), "Finding the Self? An Event-Related fMRI Study," *Journal of Cognitive Neuroscience* 14(5): 785–94.
20. Northoff, G. et al. (2006), "Self-Referential Processing in Our Brain—A Meta-Analysis of Imaging Studies on the Self," *NeuroImage* 31: 440–57.
21. Chiu, P. H. et al. (2008), "Self Responses along Cingulate Cortex Reveal Quantitative Neural Phenotype for High-Functioning Autism," *Neuron* 57(3): 463–73.
22. Moran, J. M. et al. (2013), "What Can the Organization of the Brain's Default Mode Network Tell Us About Self-Knowledge?" *Frontiers in Human Neuroscience* 7, article 391.
23. Cavanna, A. E. and Trimble, M. R. (2006), "The Precuneus: A Review of Its Functional Anatomy and Behavioural Correlates," *Brain* 129: 564–83.

24. Brewer, J. et al. (2013), "What about the 'Self' Is Processed in the Posterior Cingulate Cortex?," *Frontiers in Human Neuroscience* 7, article 647.

25. Carhart-Harris, R. L. et al. (2012), "Neural Correlates of the Psychedelic State as Determined by fMRI Studies with Psilocybin," *Proceedings of the National Academy of Sciences* 109(6): 2138–43.

26. McKenna, T., *Food of the Gods: A Radical History of Plants, Drugs and Human Evolution* (Rider, 1999).

27. Raichle, M. et al. (2001), "A Default Mode of Brain Function," *Proceedings of the National Academy of Sciences* 98(2): 676–82.

28. Sampasadaniya Sutta, Dīgha Nikāya, 28.

29. Farb, N. et al. (2007), "Attending to the Present: Mindfulness Meditation Reveals Distinct Neural Modes of Self-Reference," *SCAN* 2(4): 313–22.

30. Brewer, J. A. et al. (2011), "Meditation Experience Is Associated with Differences in Default Mode Network Activity and Connectivity," *Proceedings of the National Academy of Sciences* 108(50): 20254–59.

31. Letters of Note: The delusion. http://www.lettersofnote.com/2011/11/delusion.html (accessed March 25, 2015).

32. Armstrong, K., *Buddha* (Phoenix, 2002).

CHAPTER 6: GOLDEN SLIPPERS

1. The Mahāvagga, 1:7.

2. Nārada Mahāthera, *The Buddha and His Teachings* (Buddhist Publication Society, 2010), p. 57.

3. Wilson, T. D. et al. (2014), "Just Think: The Challenges of the Disengaged Mind," *Science* 345(6192): 75–77.

4. Raichle, M. E. et al. (2001), "A Default Mode of Brain Function," *Proceedings of the National Academy of Sciences* 98(2): 676–82.

5. Buckner, R. L. et al. (2008), "The Brain's Default Network: Anatomy, Function, and Relevance to Disease," *Annals of the New York Academy of Science* 1124: 1–38.

6. Kane, M. J. et al. (2007), "For Whom the Mind Wanders, and When: An Experience-Sampling Study of Working Memory and Executive Control in Daily Life," *Psychological Science* 18(7): 614–21.

7. Killingsworth, M. A. and Gilbert, D. T. (2010), "A Wandering Mind Is an Unhappy Mind," *Science* 330: 932.

8. Franklin, M. S. et al. (2013), "The Silver Lining of a Mind in the Clouds: Interesting Musings Are Associated with Positive Mood While Mind-Wandering," *Frontiers in Psychology* 4: 583.

9. The Dhammapada, translated by Easwaran, E. (Nilgiri, 2007), verse 1.

10. Ressler, K. and Mayberg, H. (2007), "Targeting Abnormal Neural Circuits in Mood and Anxiety Disorders: From the Laboratory to the Clinic," *Nature Neuroscience* 10: 1116–24.

11. Sheline, Y. I. et al. (2009), "The Default Mode Network and Self-Referential Processes in Depression," *Proceedings of the National Academy of Sciences* 106(6): 1942–47.

12. Nolen-Hoeksema, S. et al. (2008), "Rethinking Rumination," *Perspectives on Psychological Science* 3: 400–24.

13. Nejad, A. B. et al. (2013), "Self-Referential Processing, Rumination, and Cortical Midline Structures in Major Depression," *Frontiers in Human Neuroscience* 7, article 666.

14. Helliwell, J., Layard, R. and Sachs, J. (eds.), *World Happiness Report 2013*, pp. 5, 43.

15. World Health Organization Secretariat (2011), "Global Burden of Mental Disorders and the Need for a Comprehensive, Coordinated Response from Health and Social Sectors at the Country Level."

16. Moussavi, S. et al. (2007), "Depression, Chronic Diseases, and Decrements in Health: Results from the World Health Surveys," *The Lancet* 370(9590): 851–58.

17. Mykletun, A. et al. (2009), "Levels of Anxiety and Depression as Predictors of Mortality: The HUNT Study," *British Journal of Psychiatry* 195: 118–25.

18. Office for National Statistics (2015), *Suicides in the United Kingdom, 2013 Registrations*.

19. Centers for Disease Control and Prevention (2014), *Mortality in the United States, 2012*.

20. National Institute for Health and Care Excellence (NICE, 2009), *Depression in Adults: The Treatment and Management of Depression in Adults*; NICE clinical guideline 90.

21. Kupfer, D. J. et al. (1992), "5-Year Outcome for Maintenance Therapies in Recurrent Depression," *Archives of General Psychiatry* 49: 769–73.

22. Lau, M. A., Segal, Z. V. and Williams, J. M. G. (2004), "Teasdale's Differential Activation Hypothesis: Implications for Mechanisms of Depressive Relapse and Suicidal Behaviour," *Behaviour Research and Therapy* 42: 1001–17.

23. Segal, Z. V., Williams, J. M. G. and Teasdale, J. D., *Mindfulness-Based Cognitive Therapy for Depression* (Guilford Press, 2nd ed., 2012).

24. Lau, M. A., Segal, Z. V. and Williams, J. M. G. (2004), "Teasdale's Differential Activation Hypothesis: Implications for Mechanisms of Depressive Relapse and Suicidal Behaviour," *Behaviour Research and Therapy* 42: 1001–17.

25. Williams, M. and Penman, D., *Mindfulness: A Practical Guide to Finding Peace in a Frantic World* (Piatkus, 2011).

26. Teasdale, J. et al. (2000), "Prevention of Relapse/Recurrence in Major Depression by Mindfulness-Based Cognitive Therapy," *Journal of Consulting and Clinical Psychology* 68(4): 615–23.

27. Piet, J. and Hougaard, E. (2011), "The Effect of Mindfulness-Based Cognitive Therapy for Prevention of Relapse in Recurrent Major Depressive Disorder: A Systematic Review and Meta-Analysis," *Clinical Psychology Review* 31: 1032–40.

28. National Institute for Health and Care Excellence (NICE, 2009), *Depression in Adults: The Treatment and Management of Depression in Adults*; NICE clinical guideline 90.

29. Kuyken, W. et al. (2015), "Effectiveness and Cost-Effectiveness of Mindfulness-Based Cognitive Therapy Compared with Maintenance Anti-Depressant Treatment in the Prevention of Depressive Relapse/Recurrence: Results of the PREVENT Randomised Controlled Trial," *The Lancet* 386: 63–73.

30. Williams, M. et al. (2014), "Mindfulness-Based Cognitive Therapy for Preventing Relapse in Recurrent Depression: A Randomized Dismantling Trial," *Journal of Consulting and Clinical Psychology* 82(2): 275–86.

31. Kuyken, W. et al. (2010), "How Does Mindfulness-Based Cognitive Therapy Work?" *Behaviour Research and Therapy* 48: 1105–12.

32. Rahman, A. et al. (2013), "Grand Challenges: Integrating Maternal Mental Health into Maternal and Child Health Programmes," *PLOS Medicine* 10(5): e1001442.

33. Edwards, V. J. et al. (2003), "Relationship Between Multiple Forms of Childhood Maltreatment and Adult Mental Health in Community Respondents: Results from the Adverse Childhood Experiences Study," *American Journal of Psychiatry* 160(8): 1453–60.

34. Heim, C. et al. (2008), "The Link Between Childhood Trauma and Depression: Insights from HPA Axis Studies In Humans," *Psychoneuroendocrinology* 33: 693–710.

35. Fair, D. et al. (2008), "The Maturing Architecture of the Brain's Default Network," *Proceedings of the National Academy of Sciences* 41: 45–57.

36. Ressler, K. and Mayberg, H. (2007), "Targeting Abnormal Neural Circuits in Mood and Anxiety Disorders: From the Laboratory to the Clinic," *Nature Neuroscience* 10: 1116–24.

37. Michalak, J. et al. (2011), "Rumination as a Predictor of Relapse in Mindfulness-Based Cognitive Therapy for Depression," *Psychology and Psychotherapy* 84: 230–36.

38. Williams, J. M. G. et al. (2008), "Mindfulness-Based Cognitive Therapy (MBCT) in Bipolar Disorder: Preliminary Evaluation of Immediate Effects on Between-Episode Functioning," *Journal of Affective Disorders* 107: 275–79.

39. Deckersbach, T. et al., *Mindfulness-Based Cognitive Therapy for Bipolar Disorder* (Guilford Press, 2014).

40. Banks, K., Newman, E. and Saleem, J. (2015), "An Overview of the Re-

search on Mindfulness-Based Interventions for Treating Symptoms of Post-Traumatic Stress Disorder: A Systematic Review," *Journal of Clinical Psychology* (doi: 10.1002/jclp.22200).

41. Lucy, H., Strauss, C. and Taylor, B. L. (2013), "The Effectiveness and Acceptability of Mindfulness-Based Therapy for Obsessive Compulsive Disorder: A Review of the Literature," *Mindfulness* 4: 375–82.

42. Kessler, R. C. et al. (2005), "Lifetime Prevalence and Age-of-Onset Distributions of DSM-IV Disorders in the National Comorbidity Survey Replication," *Archives of General Psychiatry* 62(6): 593–602.

43. Kuyken, W. et al. (2013), "Effectiveness of the Mindfulness in Schools Programme: Non-Randomised Controlled Feasibility Study," *British Journal of Psychiatry* 203(2): 126–31.

44. Milton, J., *Paradise Lost* (Dover Publications, 2005).

45. The Mahāvagga, 1:7.

46. Ibid, 1:11.

47. Williams, M. and Penman, D., *Mindfulness: A Practical Guide to Finding Peace in a Frantic World* (Piatkus, 2011).

CHAPTER 7: FIRE WORSHIPPERS

1. Saṃyutta Nikāya, 35:28.

2. WHO *Report on the Global Tobacco Epidemic*, 2011.

3. Rehm, J. et al. (2009), "Global Burden of Disease and Injury and Economic Cost Attributable to Alcohol Use and Alcohol-Use Disorders," *The Lancet* 373: 2223.

4. Ipsos MORI, "Young People Omnibus 2013: A Research Study on Gambling Amongst 11–16 Year Olds on Behalf of the National Lottery Commission."

5. National Centre for Social Research (2011), *British Gambling Prevalence Survey 2010*.

6. The Bhagavad Gita, translated by Easwaran, E. (Nilgiri Press, 2007), Introduction, p. 55.

7. Brewer, J. et al. (2011), "Meditation Experience Is Associated with Differences in Default Mode Network Activity and Connectivity," *Proceedings of the National Academy of Sciences* 108(50): 20254–59.

8. Creswell, J. D. et al. (2016), "Alterations in Resting State Functional Connectivity Link Mindfulness Meditation with Reduced Interleukin-6: A Randomized Controlled Trial," *Biological Psychiatry*; published online January 29, 2016 (doi: http://dx.doi.org/10.1016/j.biopsych.2016.01.008).

9. Garavan, H. et al. (2000), "Cue-Induced Cocaine Craving: Neuroanatomical Specificity for Drug Users and Drug Stimuli," *American Journal of Psychiatry* 157: 1789–98.

10. Denton, D. et al. (1999), "Neuroimaging of Genesis and Satiation of Thirst and an Interoceptor-Driven Theory of Origins of Primary Con-

sciousness," *Proceedings of the National Academy of Sciences* 96(9): 5304–09.

11. Jarraya, B. et al. (2010), "Disruption of Cigarette Smoking Addiction after Posterior Cingulate Damage: Case Report," *Journal of Neurosurgery* 113: 1219–21.

12. Brewer, J. et al. (2013), "What about the 'Self' Is Processed in the Posterior Cingulate Cortex?," *Frontiers in Human Neuroscience* 7, article 647.

13. Kühn, S. and Gallinat, J. (2011), "Common Biology of Craving across Legal and Illegal Drugs—A Quantitative Meta-Analysis of Cue-Reactivity for Brain Response," *European Journal of Neuroscience* 33: 1318–26.

14. Brewer, J. et al. (2011), "Mindfulness Training for Smoking Cessation: Results from a Randomized Controlled Trial," *Drug and Alcohol Dependence* 119: 72–80.

15. Elwafi, H. M. et al. (2013), "Mindfulness Training for Smoking Cessation: Moderation of the Relationship Between Craving and Cigarette Use," *Drug and Alcohol Dependence* 130: 222–29.

16. National Health Interview Survey, United States, 2010.

17. McLellan, A. T. et al. (2000), "Drug Dependence, a Chronic Medical Illness: Implications for Treatment, Insurance, and Outcomes Evaluation," *JAMA* 284(13): 1689–95.

18. Ferguson, S. G. and Shiffman, S. (2009), "The Relevance and Treatment of Cue-Induced Cravings in Tobacco Dependence," *Journal of Substance Abuse Treatment* 36: 235–43.

19. Bowen, S. et al. (2014), "Relative Efficacy of Mindfulness-Based Relapse Prevention, Standard Relapse Prevention, and Treatment as Usual for Substance Use Disorders," *JAMA Psychiatry* 71(5): 547–56.

20. Bowen, S., *Mindfulness-Based Relapse Prevention for Addictive Behaviours: A Clinician's Guide* (Guilford Press, 2010).

21. Goldstein, R. and Volkow, N. (2011), "Dysfunction of the Prefrontal Cortex in Addiction: Neuroimaging Findings and Clinical Implications," *Nature Reviews: Neuroscience* 12: 652–69.

22. Tang, Y-Y. et al. (2013), "Brief Meditation Training Induces Smoking Reduction," *Proceedings of the National Academy of Sciences* 110(34): 13971–75.

23. Bowen, S. et al. (2007), "The Role of Thought Suppression in the Relation Between Mindfulness Meditation and Alcohol Use," *Addictive Behaviors* 32(10): 2324–28.

CHAPTER 8: A DRUNK ELEPHANT

1. Cullahaṁsa-Jātaka, *The Jātaka*, Book XXI, no. 533.

2. *Vinaya Texts Part III*, translated by Rhys Davids, T. W. and Oldenberg, H. (Kindle edition), p. 505.

3. Armstrong, K., *Buddha* (Phoenix, 2002), p. 150.
4. Common Ground: Solutions for Reducing the Human, Economic and Conservation Costs of Human Wildlife Conflict (World Wide Fund for Nature, 2008), p. 36.
5. The Dhammapada, translated by Easwaran, E. (The Blue Mountain Center of Meditation, 2007), verses 320–21.
6. Ibid., verses 326–27.
7. Lazar, S. W. et al. (2005), "Meditation Experience Is Associated with Increased Cortical Thickness," *Neuroreport* 16(17): 1893–97.
8. Fox, K. C. R. et al. (2014), "Is Meditation Associated with Altered Brain Structure? A Systematic Review and Meta-Analysis of Morphometric Neuroimaging in Meditation Practitioners," *Neuroscience and Behavioral Reviews* 43: 48–73.
9. Hölzel, B. K. et al. (2011), "Meditation Practice Leads to Increases in Regional Brain Gray Matter Density," *Psychiatry Research: Neuroimaging* 191: 36–43.
10. Davidson, R. J. (2000), "Affective Style, Psychopathology, and Resilience: Brain Mechanisms and Plasticity," *American Psychologist* 55(11): 1196–1214.
11. Davidson, R. J. et al. (2003), "Alterations in Brain and Immune Function Produced by Mindfulness Meditation," *Psychosomatic Medicine* 65: 564–70.
12. Dusek, J. et al. (2008), "Genomic Counter-Stress Changes Induced by the Relaxation Response," *PLOS ONE* 3: e2576.
13. Bhasin, M. K. et al. (2013), "Relaxation Response Induces Temporal Transcriptome Changes in Energy Metabolism, Insulin Secretion and Inflammatory Pathways," *PLOS ONE* 8: e62817.
14. Epel, E. S. et al. (2004), "Accelerated Telomere Shortening in Response to Life Stress," *Proceedings of the National Academy of Sciences* 101 (49): 17, 312–17.
15. Epel, E. S. et al. (2006), "Cell Aging in Relation to Stress Arousal and Cardiovascular Disease Risk Factors," *Psychoneuroendocrinology* 31: 277–87.
16. Jacobs, T. L. et al. (2011), "Intensive Meditation Training, Immune Cell Telomerase Activity, and Psychological Mediators," *Psychoneuroendocrinology* 36: 664–81.
17. Chiesa, A., Serretti, A. and Jakobsen, J. C. (2013), "Mindfulness: Top-Down or Bottom-Up Emotion Regulation Strategy?" *Clinical Psychology Review* 33: 82–96.
18. Tang, Y-Y, Hölzel, B. K. and Posner, M. I. (2015), "The Neuroscience of Mindfulness Meditation," *Nature Reviews: Neuroscience* 16: 213–25.
19. Gross, J. J. and John, O. P. (2003), "Individual Differences in Two Emotion Regulation Processes: Implications for Affect, Relationships,

and Well-Being," *Journal of Personality and Social Psychology* 85(2): 348–62.

20. Wegner, D. M. (1994), "Ironic Processes of Mental Control," *Psychological Review* 101: 163–206.

21. Dostoevsky, F. M., *Winter Notes on Summer Impressions* (Northwestern University Press, 1997).

22. Lutz, J. et al. (2014), "Mindfulness and Emotion Regulation—an fMRI Study," *Social Cognitive and Affective Neuroscience* 9(6): 776–85.

23. Micah, A. et al. (2012), "Cognitive-Affective Neural Plasticity Following Active-Controlled Mindfulness Intervention," *The Journal of Neuroscience* 32(44): 15601–10.

24. Desbordes, G. et al. (2012), "Effects of Mindful-Attention and Compassion Meditation Training on Amygdala Response to Emotional Stimuli in an Ordinary, Non-Meditative State," *Frontiers in Human Neuroscience* 6, article 292.

25. Schuyler, B. S. et al. (2014), "Temporal Dynamics of Emotional Responding: Amygdala Recovery Predicts Emotional Traits," *Social Cognitive and Affective Neuroscience* 9(2): 176–81.

26. Hölzel, B. K. et al. (2013), "Neural Mechanisms of Symptom Improvements in Generalized Anxiety Disorder Following Mindfulness Training," *NeuroImage: Clinical* 2: 448–58.

27. Chiesa, A., Serretti, A. and Jakobsen, J. C. (2013), "Mindfulness: Top-Down or Bottom-Up Emotion Regulation Strategy?" *Clinical Psychology Review* 33: 82–96.

28. Tang, Y-Y, Hölzel, B. K. and Posner, M. I. (2015), "The Neuroscience of Mindfulness Meditation," *Nature Reviews: Neuroscience* 16: 213–25.

29. Hölzel, B. K. et al. (2008), "Investigation of Mindfulness Meditation Practitioners with Voxel-Based Morphometry," *Social Cognitive and Affective Neuroscience* 3(1): 55–61.

30. Hölzel, B. K. et al. (2011), "Meditation Practice Leads to Increases in Regional Brain Gray Matter Density," *Psychiatry Research: Neuroimaging* 191: 36–43.

31. Christopher, M. S., Christopher, V. and Charoensuk, S. (2009), "Assessing 'Western' Mindfulness among Thai Theravada Buddhist Monks," *Mental Health, Religion & Culture* 12(3): 303–14.

32. Grossman, P. (2011), "Defining Mindfulness by How Poorly I Think I Pay Attention During Everyday Awareness and Other Intractable Problems for Psychology's (Re)Invention of Mindfulness: Comment on Brown et al.," *Psychological Assessment* 23(4): 1034–40.

33. Levinson, D. B. et al. (2014), "A Mind You Can Count On: Validating Breath Counting as a Behavioural Measure of Mindfulness," *Frontiers in Psychology* 5, article 1202.

34. Desbordes, G. et al. (2015), "Moving Beyond Mindfulness: Defining

Equanimity as an Outcome Measure in Meditation and Contemplative Research," *Mindfulness* 6: 356–72.

35. Davidson, R. J. et al. (2003), "Alterations in Brain and Immune Function Produced by Mindfulness Meditation," *Psychosomatic Medicine* 65: 564–70.

36. Davidson, R. J. (2000), "Affective Style, Psychopathology, and Resilience: Brain Mechanisms and Plasticity," *American Psychologist* 55(11): 1196–1214.

37. Segerstrom, S. C. and Miller, G. E. (2004), "Psychological Stress and the Human Immune System: A Meta-Analytic Study of 30 Years of Enquiry," *Psychological Bulletin* 130(4): 601–30.

38. Diener, E. and Chan, M. Y. (2011), "Happy People Live Longer: Subjective Well-Being Contributes to Health and Longevity," *Applied Psychology: Health and Well-Being* 3(1): 1–43.

39. Post, S. G. (2005), "Altruism, Happiness, and Health: It's Good to Be Good," *International Journal of Behavioural Medicine* 12(2): 66–77.

40. Zalli, A. et al. (2014), "Shorter Telomeres with High Telomerase Activity Are Associated with Raised Allostatic Load and Impoverished Psychosocial Resources," *Proceedings of the National Academy of Sciences* 111(12): 4519–24.

41. Miller, T. Q. et al. (2011), "A Meta-Analytic Review of Research on Hostility and Physical Health," *Psychological Bulletin* 119(2): 322–48.

42. Lim, D., Condon, P. and DeSteno, D. (2015), "Mindfulness and Compassion: An Examination of Mechanism and Scalability," *PLOS ONE* 10(2): e0118221.

43. Hoffman, S. G., Grossman, P. and Hinton, D. E. (2011), "Loving-Kindness and Compassion Meditation: Potential for Psychological Interventions," *Clinical Psychology Review* 31(7): 1126–32.

44. Kuyken, W. et al. (2010), "How Does Mindfulness-Based Cognitive Therapy Work?," *Behaviour Research and Therapy* 48: 1105–12.

45. Lutz, A. et al. (2008), "Regulation of the Neural Circuitry of Emotion by Compassion Meditation: Effects of Meditative Expertise," *PLOS ONE* 3(3): e1897.

46. Meng-Wu Lecture by Richard Davidson at the Center for Compassion and Altruism Research and Education (CCARE) at Stanford School of Medicine, October 2, 2012.

47. Lutz, A. et al. (2009), "BOLD Signal in Insula Is Differentially Related to Cardiac Function During Compassion Meditation in Experts vs. Novices," *NeuroImage* 47(3): 1038–46.

48. Lim, D., Condon, P. and DeSteno, D. (2015), "Mindfulness and Compassion: An Examination of Mechanism and Scalability," *PLOS ONE* 10(2): e0118221.

49. The Mahāvagga, VIII, 26.1–8.

50. Hoffman, S. G. et al. (2011), "Loving-Kindness and Compassion Meditation: Potential for Psychological Interventions," *Clinical Psychology Review* 31(7): 1126–32.

51. McCall, C. et al. (2014), "Compassion Meditators Show Less Anger, Less Punishment, and More Compensation to Victims in Response to Fairness Violations," *Frontiers in Behavioral Neuroscience* 8, article 424.

CHAPTER 9: THE FALL

1. Uher, R. (2009), "The Role of Genetic Variation in the Causation of Mental Illness: An Evolution-Informed Framework," *Molecular Psychiatry* 14: 1072–82.

2. Kessler, R. C. et al. (2005), "Lifetime Prevalence and Age-of-Onset Distributions of DSM-IV Disorders in the National Comorbidity Survey Replication," *Archives of General Psychiatry* 62: 593–602.

3. Andrews, P., Poulton, R. and Skoog, I. (2005), "Lifetime Risk for Depression: Restricted to a Minority or Waiting for Most?," *British Journal of Psychiatry* 187: 495–96.

4. Moffitt, T. E. et al. (2010), "How Common Are Common Mental Disorders? Evidence that the Lifetime Prevalence Rates Are Doubled by Prospective Versus Retrospective Ascertainment," *Psychological Medicine* 40: 899–909.

5. Uher, R. (2009), "The Role of Genetic Variation in the Causation of Mental Illness."

6. Cross-Disorder Group of the Psychiatric Genomics Consortium (2013), "Identification of Risk Loci with Shared Effects on Five Major Psychiatric Disorders: A Genome-Wide Analysis," *The Lancet* 381: 1371–79.

7. Genetics of Personality Consortium (2015), "Meta-Analysis of Genome-Wide Association Studies for Neuroticism, and the Polygenic Association with Major Depressive Disorder," *JAMA Psychiatry*; published online May 20, 2015 (doi:10.1001/jamapsychiatry.2015.0554).

8. Seneca, "On Tranquillity of Mind," in: *Dialogues and Letters*, translated by Costa, C. D. N. (Penguin Classics, 1997).

9. Kyaga, S. et al. (2011), "Creativity and Mental Disorder: Family Study of 300,000 People with Severe Mental Disorder," *The British Journal of Psychiatry* 199: 373–79.

10. Nettle, D. and Clegg, H. (2006), "Schizotypy, Creativity and Mating Success in Humans," *Proceedings of the Royal Society B* 273: 611–15.

11. Power, R. A. et al. (2013), "Fecundity of Patients with Schizophrenia, Autism, Bipolar Disorder, Depression, Anorexia Nervosa, or Substance Abuse vs Their Unaffected Siblings," *JAMA Psychiatry* 70(1): 22–30.

12. Uher, R. (2009), "The Role of Genetic Variation in the Causation of Mental Illness."

13. Yee, C. M., Javitt, D. C. and Miller, G. A. (2015), "Replacing *DSM* Categorical Analyses with Dimensional Analyses in Psychiatry Research: The Research Domain Criteria Initiative," *JAMA Psychiatry* 72(12): 1159–60.

14. Adam, D. (2013), "Mental Health: On the Spectrum," *Nature* 496: 416–18.

15. Bebbington, P. E. et al. (2013), "The Structure of Paranoia in the General Population," *The British Journal of Psychiatry* 202(6): 419–27.

16. Carroll, L., *Alice in Wonderland and Through the Looking Glass* (Pan, 1947).

17. Almécija, S. et al. (2013), "The Femur of *Orrorin tugenensis* Exhibits Morphometric Affinities with Both Miocene Apes and Later Hominins," *Nature Communications* 4: article 2888 (doi:10.1038/ncomms3888).

18. Richmond, B. G. and Jungers, W. L. (2008), "*Orrorin tugenensis* Femoral Morphology and The Origin of Hominin Bipedalism," *Science* 319: 1662–65.

19. Dunbar, R., *Human Evolution* (Penguin, 2014), pp. 109–25.

20. Corballis, M. C., *The Recursive Mind: The Origins of Human Language, Thought, and Civilization* (Princeton University Press, 2011), pp. 55–79.

21. Savage-Rumbaugh, S., *Kanzi: The Ape at the Brink of the Human Mind* (Wiley, 1996).

22. Roberts, A. I., Vick, S.-J. and Buchanan-Smith, H. J. (2012), "Usage and Comprehension of Manual Gestures in Wild Chimpanzees," *Animal Behaviour* 84: 459–70.

23. Rizzolatti, G. et al. (1996), "Premotor Cortex and the Recognition of Motor Actions," *Cognitive Brain Research* 3: 131–41.

24. Corballis, M. C. (2013), "Wandering Tales: Evolutionary Origins of Mental Time Travel and Language," *Frontiers in Psychology* 4, article 485.

25. Sheline, Y. I. et al. (2009), "The Default Mode Network and Self-Referential Processes in Depression," *Proceedings of the National Academy of Sciences* 106(6): 1942–47.

26. Nejad, A. B., Fossati, P. and Lemogne, C. (2013), "Self-Referential Processing, Rumination, and Cortical Midline Structures in Major Depression," *Frontiers in Human Neuroscience* 7, article 666.

27. Marchetti, I. et al. (2012), "The Default Mode Network and Recurrent Depression: A Neurobiological Model of Cognitive Risk Factors," *Neuropsychology Review* 22(3): 229–51.

28. Broyd, S. J. (2008), "Default-Mode Brain Dysfunction in Mental Disorders: A Systematic Review," *Neuroscience and Biobehavioral Reviews* 33(3): 279–96.

29. Sun, L. et al. (2012), "Abnormal Functional Connectivity between the Anterior Cingulate and the Default Mode Network in Drug-Naive Boys

with Attention Deficit Hyperactivity Disorder," *Psychiatry Research: Neuroimaging* 201: 120–27.

30. Whitfield, S. et al. (2009), "Hyperactivity and Hyperconnectivity of the Default Network in Schizophrenia and in First-Degree Relatives of Persons with Schizophrenia," *Proceedings of the National Academy of Sciences* 106(4): 1279–84.

31. Chai, X. J. et al. (2011), "Abnormal Medial Prefrontal Cortex Resting-State Connectivity in Bipolar Disorder and Schizophrenia," *Neuropsychopharmacology* 36: 2009–17.

32. Whitfield-Gabrieli, S. and Ford, J. M. (2012), "Default Mode Network Activity and Connectivity in Psychopathology," *Annual Reviews: Clinical Psychology* 8: 49–76.

33. Kyaga, S. et al. (2011), "Creativity and Mental Disorder: Family Study of 300,000 People with Severe Mental Disorder."

34. Kyaga, S. et al. (2013), "Mental Illness, Suicide and Creativity: 40-Year Prospective Total Population Study," *Journal of Psychiatric Research* 47(1): 83–90.

35. Dunbar, R., *Human Evolution* (Pelican, 2014), p. 65.

36. Dunbar, R. (1998), "The Social Brain Hypothesis," *Brain* 9(10): 178–90.

37. Hurdiel, R. et al. (2012), "Field Study of Sleep and Functional Impairments in Solo Sailing Races," *Sleep and Biological Rhythms* 10(4): 270–77.

38. Grassian, S. (1983), "Psychopathological Effects of Solitary Confinement," *American Journal of Psychiatry* 140(11): 1450–54.

39. Gen. 3:7 (AV).

40. Anonymous, *The Cloud of Unknowing*, translated by Spearing, A. C. (Penguin Classics, 2001).

CHAPTER 10: WONDERFUL AND MARVELOUS

1. Majjhima Nikāya, 123.

2. Tang, Y-Y, Hölzel, B. and Posner, I. P. (2015), "The Neuroscience of Mindfulness Meditation," *Nature Reviews: Neuroscience* 16: 213–25.

3. Lazar, S. W. et al. (2005), Meditation Experience Is Associated with Increased Cortical Thickness," *Neuroreport* 16(17): 27–33.

4. Luders, E. et al. (2009), "The Underlying Anatomical Correlates of Long-Term Meditation," *NeuroImage* 45(3): 672–78.

5. Fox, K. C. R. et al. (2014), "Is Meditation Associated with Altered Brain Structure? A Systematic Review and Meta-Analysis of Morphometric Neuroimaging in Meditation Practitioners," *Neuroscience & Biobehavioral Reviews* 43: 48–73.

6. Burgess, P. W. et al., "The Gateway Hypothesis of Rostral Prefrontal Cortex (Area 10) Function," in: Duncan, J., Phillips, L. and McLeod, P.

(eds.), *Measuring the Mind: Speed, Control, and Age* (Oxford University Press, 2005), pp. 217–48.

7. Fox, K. C. R. et al. (2014), "Is Meditation Associated with Altered Brain Structure?: A Systematic Review and Meta-Analysis of Morphometric Neuroimaging in Meditation Practitioners," *Neuroscience and Biobehavioral Reviews* 43: 48–73.

8. Neubert, F-X. et al. (2014), "Comparison of Human Ventral Frontal Cortex Areas for Cognitive Control and Language with Areas in Monkey Frontal Cortex," *Neuron* 81: 700–13.

9. Baird, B. et al. (2013), "Medial and Lateral Networks in Anterior Prefrontal Cortex Support Metacognitive Ability for Memory and Perception," *The Journal of Neuroscience* 33(42): 16657–65.

10. Burgess, P. W. et al., "The Gateway Hypothesis of Rostral Prefrontal Cortex (Area 10) Function."

11. Zhou, Y. et al. (2015), "The Selective Impairment of Resting-State Functional Connectivity of the Lateral Subregion of the Frontal Pole in Schizophrenia," *PLOS ONE* 10(3): e0119176.

12. Liu, H. et al., "Connectivity-Based Parcellation of the Human Frontal Pole with Diffusion Tensor Imaging," *The Journal of Neuroscience* 33(16): 6782–90.

13. Koenigs, M. and Grafman, J. (2009), "The Functional Neuroanatomy of Depression: Distinct Roles for Ventromedial and Dorsolateral Prefrontal Cortex," *Behavioural and Brain Research* 201(2): 239–43.

14. Farb, N. A. S. et al. (2011), "Mood-Linked Responses in Medial Prefrontal Cortex Predict Relapse in Patients with Recurrent Unipolar Depression," *Biological Psychiatry* 70(4): 366–72.

15. Hooley, J. M. et al. (2005), "Activation in Dorsolateral Prefrontal Cortex in Response to Maternal Criticism and Praise in Recovered Depressed and Healthy Control Participants," *Biological Psychiatry* 57(7): 809–12.

16. Teasdale, J. D. et al. (2002), "Metacognitive Awareness and Prevention of Relapse in Depression: Empirical Evidence," *Journal of Consulting and Clinical Psychology* 70(2): 175–87.

17. Tang, Y-Y, Hölzel, B. and Posner, I. P. (2015), "The Neuroscience of Mindfulness Meditation."

18. Freeman, D. and Freeman, J., "Forget the Headlines, Schizophrenia Is More Common Than You Might Think," theguardian.com, November 15, 2013.

19. Adam, D. (2013), "Mental Health: On the Spectrum," *Nature* 496: 416–18.

20. Craig, (Bud) A. D. (2009), "How Do You Feel—Now? The Anterior Insula and Human Awareness," *Nature Reviews: Neuroscience* 10: 59–70.

21. Fox, K. C. R. et al. (2014), "Is Meditation Associated with Altered Brain Structure?"

22. Tang, Y-Y, Hölzel, B. and Posner, I. P. (2015), "The Neuroscience of Mindfulness Meditation."

23. Hasenkamp, W. et al. (2012), "Mind Wandering and Attention During Focused Meditation: A Fine-Grained Temporal Analysis of Fluctuating Cognitive States," *NeuroImage* 59: 750–60.

24. Picard, F. and Craig, A. D. (2009), "Ecstatic Epileptic Seizures: A Potential Window on the Neural Basis of Self-Awareness," *Epilepsy & Behavior* 16: 539–46.

25. Picard, F., Scavarda, D. and Bartolomei, F. (2013), "Induction of a Sense of Bliss by Electrical Stimulation of the Anterior Insula," *Cortex* 49(10): 2935–37.

26. Anonymous, *The Cloud of Unknowing*, translated by Spearing, A. C. (Penguin Classics, 2001).

27. Majjhima Nikāya, 36.

28. Picard, F. and Kurth, F. (2014), "Ictal Alterations of Consciousness during Ecstatic Seizures," *Epilepsy & Behavior* 30: 58–61.

29. Harris, S., *Waking Up: A Guide to Spirituality Without Religion* (Simon & Schuster, 2014).

30. Fox, K. C. R. et al. (2013), "Dreaming As Mind-Wandering: Evidence from Functional Neuroimaging and First-Person Content Reports," *Frontiers in Human Neuroscience* 7, article 412.

31. Strathern, P., *Mendeleyev's Dream: The Quest for the Elements* (St. Martin's Press, 2000).

32. Loewi, O. (1960), "An Autobiographical Sketch," *Perspectives in Biological Medicine* 4: 1–25.

33. Thich Nhat Hanh, *The Heart of the Buddha's Teaching* (Ebury Publishing, 1999).

CHAPTER 11: MIND MIRRORS

1. "Confidence in Our Own Ability," a Dhamma talk given by Ajahn Amaro at Amaravati Buddhist Monastery on August 7, 2015 (http://bit.ly/1MaBPDa).

2. Anguttara Nikāya, 3.65.

3. Harrington, A. and Zajonc, A. (eds.), *The Dalai Lama at MIT* (Harvard University Press, 2008), p. 63.

4. His Holiness the Dalai Lama, *The Universe in a Single Atom: How Science and Spirituality Can Serve Our World* (Abacus, 2005).

5. Henderson, B., "Open Letter to Kansas School Board" (http://www.venganza.org/about/open-letter/).

6. Kuyken, W. et al. (2015), "Effectiveness and Cost-Effectiveness of Mindfulness-Based Cognitive Therapy Compared with Maintenance

Anti-Depressant Treatment in the Prevention of Depressive Relapse/Recurrence: Results of the PREVENT Randomised Controlled Trial," *The Lancet* 386: 63–73.

7. Ong, J. C. et al. (2014), "A Randomized Controlled Trial of Mindfulness Meditation for Chronic Insomnia," *Sleep* 37(9): 1553–63.

8. Williams, J. M. G. et al. (2008), "Mindfulness-Based Cognitive Therapy (MBCT) in Bipolar Disorder: Preliminary Evaluation of Immediate Effects on Between-Episode Functioning," *Journal of Affective Disorders* 107: 275–79.

9. Chadwick, P. (2014), "Mindfulness for Psychosis," *The British Journal of Psychiatry* 204(5): 333–34.

10. Deckersbach, T. et al., *Mindfulness-Based Cognitive Therapy for Bipolar Disorder* (Guilford Press, 2014).

11. Dyga, K. and Stupak, R. (2015), "Meditation and Psychosis: Trigger or Cure?," *Archives of Psychiatry and Psychotherapy* 3: 48–58.

12. The Varieties of Contemplative Experience Study (cheetahhouse.org).

13. Gotink, R. A. et al. (2015), "Standardised Mindfulness-Based Interventions in Healthcare: An Overview of Systematic Reviews and Meta-Analyses of RCTs," *PLOS ONE* 10(4): e0124344.

14. Goyal, M. et al. (2014), "Meditation Programs for Psychological Stress and Well-Being: A Systematic Review and Meta-Analysis," *JAMA Internal Medicine* 174(3): 357–68.

15. Zeidan, F. et al. (2010), "Effects of Brief and Sham Mindfulness Meditation on Mood and Cardiovascular Variables," *The Journal of Alternative and Complementary Medicine* 16(8): 867–73.

16. Killingsworth, M. A. and Gilbert, D. T. (2010), "A Wandering Mind Is an Unhappy Mind," *Science* 330: 932.

17. Bowen, S. et al. (2014), "Relative Efficacy of Mindfulness-Based Relapse Prevention, Standard Relapse Prevention, and Treatment as Usual for Substance Use Disorders," *JAMA Psychiatry* 71(5): 547.

18. Sedlmeier, P. et al. (2012), "The Psychological Effects of Meditation: A Meta-Analysis," *Psychological Bulletin* 138(6): 1139–71.

19. Frederick, S. (2005), "Cognitive Reflection and Decision Making," *The Journal of Economic Perspectives* 19(4): 25–42.

20. Kahneman, D., *Thinking, Fast and Slow* (Penguin Books, 2012).

21. Brefczynski-Lewis, J. A. et al. (2007), "Neural Correlates of Attentional Expertise in Long-Term Meditation Practitioners," *Proceedings of the National Academy of Sciences* 104(27): 11483–88.

22. Slagter, H. A. et al. (2007), "Mental Training Affects Distribution of Limited Brain Resources," *PLOS Biology* 5(6): e138.

23. Lutz, A. et al. (2009), "Mental Training Enhances Attentional Stability: Neural and Behavioral Evidence," *The Journal of Neuroscience* 29: 13418–27.

24. MacCoon, D. G. et al. (2014), "No Sustained Attention Differences in a Longitudinal Randomized Trial Comparing Mindfulness Based Stress Reduction Versus Active Control," *PLOS ONE* 9(6): e97551.

25. Chiesa, A., Calati, R. and Serretti, A. (2011), "Does Mindfulness Training Improve Cognitive Abilities?," *Clinical Psychology Review* 31: 449–64.

26. Sai Sun et al. (2015), "Calm and Smart? A Selective Review of Meditation Effects on Decision Making," *Frontiers in Psychology* (http://dx.doi.org/10.3389/fpsyg.2015.01059).

27. "The Large Hadron Collider" (supplement), *Guardian*, June 30, 2008.

28. Eagleman, D., *Incognito: The Secret Lives of the Brain* (Canongate, 2011).

29. Rosario, M. R. (2005), "Training, Maturation and Genetic Influences on the Development of Executive Attention," *Proceedings of the National Academy of Sciences* 102(41): 14931–36.

30. James, W. (1890), *The Principles of Psychology* (Classics of Psychiatry and Behavioral Sciences Library, 1988).

31. Visser, S. N. et al. (2014), "Trends in the Parent-Report of Health Care Provider-Diagnosed and Medicated Attention-Deficit/Hyperactivity Disorder: United States, 2003–2011," *Journal of the American Academy of Child & Adolescent Psychiatry* 53(1): 34–46.

32. van de Weijer-Bergsma, E. et al. (2012), "The Effectiveness of Mindfulness Training on Behavioral Problems and Attentional Functioning in Adolescents with ADHD," *Journal of Child and Family Studies* 21: 775–87.

33. Buckner, R. L. et al. (2005), "Molecular, Structural, and Functional Characterization of Alzheimer's Disease: Evidence for a Relationship Between Default Activity, Amyloid, and Memory," *Journal of Neuroscience* 25: 7709–17.

34. Bero, A. W. et al. (2011), "Neuronal Activity Regulates the Regional Vulnerability to Amyloid-Deposition," *Nature Neuroscience* 14(6): 750–56.

35. Luders, E. (2014), "Exploring Age-Related Brain Degeneration in Meditation Practitioners," *Annals of the New York Academy of Sciences* 1307: 82–88.

36. Mozzafarian, D. (2015), "The 2015 US Dietary Guidelines: Lifting the Ban on Total Dietary Fat," *JAMA* 313(24): 2421–22.

37. Garrison, K. A. et al. (2013), "Effortless Awareness: Using Real Time Neurofeedback to Investigate Correlates of Posterior Cingulate Cortex Activity in Meditators' Self Report," *Frontiers in Human Neuroscience* 7: 440.

38. Garrison, K. A. et al. (2013), "Real-time fMRI Links Subjective Experi-

ence with Brain Activity During Focused Attention," *NeuroImage* 81: 110–18.

39. Amaro, A. (2015), "A Holistic Mindfulness," *Mindfulness* 6(1): 63–73.

CHAPTER 12: THE DEATHLESS REALM

1. *Buddhist Birth-Stories or Jataka Tales, volume 1,* translated by Rhys Davids, T. W. (Kindle edition, 2015).
2. Xuanzang, *The Great Tang Dynasty Record of the Western Regions* (Numata Center for Buddhist Translation and Research, 1998).
3. Allen, C., *Ashoka: The Search for India's Lost Emperor* (Abacus, 2013), p. 394.
4. Western Buddhist Teachers Conference (Dharamsala, India, March 1993), The Meridian Trust Tibetan Cultural Film Archive, http://me ridian-trust.org/video/the-western-buddhist-teachers-conference-with -h-h-the-dalai-lama-3-of-8/4.
5. Sumedho, A., *The Anthology: Volume 4, The Sound of Silence* (Amaravati Publications, 2014), p. 195.
6. Saṃyutta Nikāya, 12.2.
7. Armstrong, K., *Buddha* (Phoenix, 2000), pp. 97–100.
8. Gal. 6:8 (AV).
9. Anguttara Nikāya, 4.77.
10. Majjhima Nikāya, 63.
11. Anguttara Nikāya, 5.57.
12. The Dhammapada, translated by Easwaran, E. (Nilgiri, 2007), verses 147–48.
13. Ibid., verses 153–54.
14. Amaro, A. (2015), "A Holistic Mindfulness," *Mindfulness* 6(1): 63–73.
15. The Dhammapada, translated by Easwaran, E. (Nilgiri, 2007), verse 21.
16. Access to Insight (http://www.accesstoinsight.org/ptf/dhamma/sagga /loka.html).
17. Lutz, A. et al. (2004), "Long-Term Meditators Self-Induce High-Amplitude Gamma Synchrony During Mental Practice," *Proceedings of the National Academy of Sciences* 101(46): 16369–73.
18. van Lommel, P. (2001), "Near-Death Experience in Survivors of Cardiac Arrest: A Prospective Study in the Netherlands," *Lancet* 358: 2039–45.
19. Borjigin, J. et al. (2013), "Surge of Neurophysiological Coherence and Connectivity in the Dying Brain," *Proceedings of the National Academy of Sciences* 110(35): 14432–37.
20. Dennis Potter's final interview, broadcast by Channel 4 on April 5, 1994 (https://vimeo.com/26503584).
21. Helliwell, J., Layard, R. and Sachs, J., *World Happiness Report 2013*, p. 81.

22. Jackson, T., *Prosperity Without Growth* (Earthscan, 2011).
23. Allen, C., *Ashoka: The Search for India's Lost Emperor* (Abacus, 2013), p. 180.
24. Pinker, S., *The Better Angels of Our Nature* (Viking Books, 2011).
25. Saṃyutta Nikāya, 46.16.
26. Dīgha Nikāya, 16.

INDEX

NOTE: Bold page numbers refer to figures.

INDEX

senses/sensations (*cont.*)
  and cycle of birth and death, 281
  dependent origination and, **282**, 283
  and evolution of humans, 295
  and fire worshippers analogy, 147
  focused attention and, 256
  Guided Meditation and, 142–43
  MBCT and, 131–32
  MBSR and, 175
  metacognition and, 225
  and mindfulness, 179, 257
  regulation of, 175, 176–77
  and Siddhārtha's death, 298
  and Three-Minute Breathing Space, 142–43
  See also specific sensation
Seven Factors of Enlightenment, 296–97
Shambhala Mountain Center: meditation retreats at, 52–53
Shinto Buddhism, 48
sickle cell anemia, 198
sickness
  and compassion, 191
  dukkha and, 3
  and early history of India, 35
  and human existence as suffering, 3
  refusal to face, 20
  and Siddhārtha's upbringing, 18–20
  Siddhārtha's views about, 191
  and terminal illness, 289–90
  and transience of life, 291
  See also mental illness; *specific sickness*
Siddhārtha Gautama
  aim/goals of, 284
  and aim of mindfulness, 268, 271
  as ascetic, 15–16, 37–39, 42, 55–56, 68, 73, 74–75, 92–93
  childhood/youth of, 16, 18, 22, 24, 35, 39, 41–42, 55, 68, 172
  and components of human existence, 99
  and cultivation of healthy mind, 219
  and cycle of birth and death, 281, 286, 292
  dart metaphor of, 80
  death of, 297–98
  dependent origination concept of, 281, 283
  Devadatta's rivalry with, 168–69
  and drunken elephant incident, 168–71, 172, 184
  elephants compared with, 170
  emotions of, 172–73

enlightenment of, 20, 28–29, 33, 34, 36, 37–39, 41, 55–56, 68, 73, 74–75, 92–93, 96–97, 227, 233, 274–76, 285–86, 292
as facing realities of life, 18–19
family background of, 16, 18, 22, 24
Fire Sermon of, 146–47, 151, 161, 165, 177
followers of, 37, 55, 59–60, 73, 74–75, 92–93, 96–97, 105, 141, 217–19
and importance of reflection, 284–85
insight views of, 281
karma views of, 281, 284, 285, 294
key locations in life of, **17**
legacy of, 13, 33
and Mara, 274–75, 292
meditation of, 120
mental crisis of, 18–20
middle way of, 68, 74–76, 150–51
and nonduality approach, 92–93
rebirth views of, 283
reputation of, 37, 141, 217
as revolutionary, 92–93
Self and, 32, 91, 93, 97, 99–100, 105, 118–19, 120, 283
and Seven Factors of Enlightenment, 296–97
spreading the word of, 141
and suffering, 34, 75, 80, 90, 97, 100, 120, 141, 146, 147, 151, 268, 271, 281, 284
walking meditation of, 122
wife and children of, 18, 19
wonderful and marvelous qualities of, 218–19, 227
and Yasa, 122, 140–41, 146
Yoga and, 37, 38
See also specific topic
sin, 7, 216, 269
Singer, Tania, 190
skepticism, 28–29, 33, 244, 255, 287
Skinner, B.F., 158–59
smoke detector analogy, 178, 179
smoking, 24, 115, 148, 154, 155–58, 159, 160, 161, 164, 267
social brain hypothesis, 212, 294–95, 296
social cohesion, 25–26, 125, **211**, 212–13, 216, 220–21, 222
solitary confinement, 123, 214
somatosensory cortex, 89, 172, 177, 224
"Song of Mahamudra" (Tilopa), 40
Soulsby, Judith, 132–33, 265–66

ABOUT THE AUTHOR

James Kingsland is a science and medical journalist with twenty-five years of experience working for publications such as *New Scientist, Nature,* and most recently the *Guardian* (UK), where he was a commissioning editor and a contributor for its Notes & Theories blog. On his own blog, Plastic Brain, he writes about neuroscience and Buddhist psychology.